PROCEEDINGS OF SYMPOSIA IN APPLIED MATHEMATICS

VOLUME 1 NON-LINEAR PROBLEMS IN MECHANICS OF CONTINUA
Edited by E. Reissner (Brown University, August 1947)

VOLUME 2 ELECTROMAGNETIC THEORY
Edited by A. H. Taub (Massachusetts Institute of Technology, July 1948)

VOLUME 3 ELASTICITY
Edited by R. V. Churchill (University of Michigan, June 1949)

VOLUME 4 FLUID DYNAMICS
Edited by M. H. Martin (University of Maryland, June 1951)

VOLUME 5 WAVE MOTION AND VIBRATION THEORY
Edited by A. E. Heins (Carnegie Institute of Technology, June 1952)

VOLUME 6 NUMERICAL ANALYSIS
Edited by J. H. Curtiss (Santa Monica City College, August 1953)

VOLUME 7 APPLIED PROBABILITY
Edited by L. A. MacColl (Polytechnic Institute of Brooklyn, April 1955)

VOLUME 8 CALCULUS OF VARIATIONS AND ITS APPLICATIONS
Edited by L. M. Graves (University of Chicago, April 1956)

VOLUME 9 ORBIT THEORY
Edited by G. Birkhoff and R. E. Langer (New York University, April 1957)

VOLUME 10 COMBINATORIAL ANALYSIS
Edited by R. Bellman and M. Hall, Jr. (Columbia University, April 1958)

VOLUME 11 NUCLEAR REACTOR THEORY
Edited by G. Birkhoff and E. P. Wigner (New York City, April 1959)

VOLUME 12 STRUCTURE OF LANGUAGE AND ITS MATHEMATICAL ASPECTS
Edited by R. Jakobson (New York City, April 1960)

VOLUME 13 HYDRODYNAMIC INSTABILITY
Edited by R. Bellman, G. Birkhoff, C. C. Lin (New York City, April 1960)

VOLUME 14 MATHEMATICAL PROBLEMS IN THE BIOLOGICAL SCIENCES
Edited by R. Bellman (New York City, April 1961)

VOLUME 15 EXPERIMENTAL ARITHMETIC, HIGH SPEED COMPUTING, AND MATHEMATICS
Edited by N. C. Metropolis, A. H. Taub, J. Todd, C. B. Tompkins (Atlantic City and Chicago, April 1962)

VOLUME 16 STOCHASTIC PROCESSES IN MATHEMATICAL PHYSICS AND ENGINEERING
Edited by R. Bellman (New York City, April 1963)

VOLUME 17 **APPLICATIONS OF NONLINEAR PARTIAL DIFFERENTIAL EQUATIONS IN MATHEMATICAL PHYSICS**
Edited by R. Finn (New York City, April 1964)

VOLUME 18 **MAGNETO-FLUID AND PLASMA DYNAMICS**
Edited by H. Grad (New York City, April 1965)

VOLUME 19 **MATHEMATICAL ASPECTS OF COMPUTER SCIENCE**
Edited by J. T. Schwartz (New York City, April 1966)

VOLUME 20 **THE INFLUENCE OF COMPUTING ON MATHEMATICAL RESEARCH AND EDUCATION**
Edited by J. P. LaSalle (University of Montana, August 1973)

VOLUME 21 **MATHEMATICAL ASPECTS OF PRODUCTION AND DISTRIBUTION OF ENERGY**
Edited by P. D. Lax (San Antonio, Texas, January 1976)

VOLUME 22 **NUMERICAL ANALYSIS**
Edited by G. H. Golub and J. Oliger (Atlanta, Georgia, January 1978)

VOLUME 23 **MODERN STATISTICS: METHODS AND APPLICATIONS**
Edited by R. V. Hogg (San Antonio, Texas, January 1980)

VOLUME 24 **GAME THEORY AND ITS APPLICATIONS**
Edited by W. F. Lucas (Biloxi, Mississippi, January 1979)

VOLUME 25 **OPERATIONS RESEARCH: MATHEMATICS AND MODELS**
Edited by S. I. Gass (Duluth, Minnesota, August 1979)

VOLUME 26 **THE MATHEMATICS OF NETWORKS**
Edited by S. A. Burr (Pittsburgh, Pennsylvania, August 1981)

AMS SHORT COURSE LECTURE NOTES
published as a subseries of
Proceedings of Symposia in Applied Mathematics

PROCEEDINGS OF SYMPOSIA
IN APPLIED MATHEMATICS
Volume 27

COMPUTED TOMOGRAPHY

AMERICAN MATHEMATICAL SOCIETY
PROVIDENCE, RHODE ISLAND

LECTURE NOTES PREPARED FOR THE
AMERICAN MATHEMATICAL SOCIETY SHORT COURSE

COMPUTED TOMOGRAPHY

HELD IN CINCINNATI, OHIO
JANUARY 11–12, 1982

EDITED BY
LAWRENCE A. SHEPP

The AMS Short Course Series is sponsored by the Society's Committee on Employment and Education Policy (CEEP). The Series is under the direction of the Short Course Advisory Subcommittee of CEEP.

Library of Congress Cataloging in Publication Data

Main entry under title:
Computed tomography.
(Proceedings of symposia in applied mathematics, ISSN 0160-7634; v. 27)
"Lecture notes prepared for the American Mathematical Society short course, computed tomography, held in Cincinnati, Ohio, January 11–12, 1982"––Verso t.p.
Includes bibliographies.
1. Tomography––Mathematics––Congresses. I. Shepp, Lawrence A., 1936– . II. American Mathematical Society. III. Series.
RC78.7.T6C6417 1982 616.07′572 82-18508
ISBN 0-8218-0033-7

1980 Mathematics Subject Classification. Primary 42B20.

Printed in the United States of America.

This volume was printed directly from copy prepared by the authors.

CONTENTS

PREFACE

Since its invention in 1972 by Godfrey N. Hounsfield, computed tomography (CT) has benefitted hundreds of thousands of patients and represents a business of over 500 million dollars annually.

The central technology in CT is unquestionably recognized to be mathematics, not only in the medical community but in the physics and engineering areas of CT as well. It is unfortunate that the mathematical community itself has less of an appreciation and understanding of this important application of mathematics. The reason for this is the usual one, that mathematicians are often content to work on questions "of intrinsic interest", and there are always plenty of these around which do not have to deal with the realities of a real-world problem such as CT.

Choosing the right level of abstraction in a mathematical model of a real-world is the key to effective applied mathematics. An engineering type trade-off must be made between the computability or analyzability of each particular model and its accuracy or relevance to the real problem. Often several models, partly contradictory, must be used.

In this collection of papers, which have been arranged in decreasing relevance or increasing abstraction, mathematicians unfamiliar with applied mathematics will be able to see a rather full spectrum of models of (both old and new areas of) CT. This experience we have gained in CT will (hopefully) be helpful to us and the reader in choosing models for other applied problems. Finally there are nice problems both relevant and "of intrinsic interest" suggested by each of the papers. Without the latter, mathematics and mathematicians would be less interesting and, by analogy with the work of Radon which no doubt appeared to be irrelevant at the time, would perhaps be less relevant.

Proceedings of Symposia in Applied Mathematics
Volume 27, 1982

SCOPE OF PURE AND APPLIED TOMOGRAPHY

L. A. Shepp

Abstract. These lecture notes for the AMS Short Course on Computed Tomography (Jan, 1982) discuss and philosophize on the mathematical basis of reconstruction CT (computed tomography). Besides standard X-ray CT, we discuss the newer problem of NMR (nuclear magnetic resonance) CT and sketch a new approach to PET (positron emission tomography) based on a theorem in mathematical statistics.

§1. Introduction

The bumper sticker: "Mathematicians Do It In Their Heads", can also be read to illustrate the frequent characterization of the mathematician as one who ignores the real world and prefers to be imbedded in axiomatics and idealized abstract models. Nevertheless, the advantages of the mathematical approach, based on a well-defined model, to a real problem are manifold: clarity, computability, elegance, to name a few. Except for classical mechanics and a few such cases, no model usually exists which is simultaneously clear, computable, and elegant, and this unfortunately also holds for tomography. It would be great if one could say that CT is all in Radon's famous 1917 paper and the world simply waited for computers to come along, but this is a gross oversimplification. In fact much work was required to adapt Radon's inversion formula to the real world of (a finite number of) imperfect measurements. To ignore the details of such adaptation is to "Do It In Your Head". I have selected the individual speakers of the course to present both the mathematical and non-mathematical approaches on an equal time basis. Each individual in the audience can judge whether he is viewing pure or applied tomography.

I will now discuss the three current CT areas, X-ray CT, NMR CT, and PET first using the mathematical model where all measurements can be performed and with no errors, and where explicit inversion will be possible. Then we will discuss the modifications in the models that are necessary if one is to take into account the limitations that the total number of measurements must be finite and that noise of various types can enter.

§2. X-ray CT

Let $f=f(x,y)$ be a continuous function on the plane supported on $x^2+y^2<1$. For any line L, let $P(L)=\int_L f ds$ where s is length measured along L. Then Radon [R] gave the explicit inversion formula for f in terms of P:

$$f(Q)=-\frac{1}{\pi}\int_0^\infty \frac{1}{q}dF_Q(q) \tag{2.1}$$

where $F_Q(q)$ is the average value of the line integral $P(L)$ for lines L at distance q from Q, i.e.

$$F_Q(q)=\frac{1}{2\pi}\int_0^{2\pi} P(L_{Q,\theta}^{q})d\theta \tag{2.2}$$

0160-7634/82/0000-0930/$02.50

where $L^q_{Q,\theta}$ is the line tangent to the circle of radius q about Q which makes angle θ with the positive x-axis. As is shown in [R], the integral in (2.1) converges as a Stieltjes integral.

Although (2.1)-(2.2) solve the problem of reconstructing f from P for the case when *all* line integrals of f are exactly known, this formula does not apply to the actual problem of real world X-ray CT where only a finite number of measurements can be made. How should the model be modified? It is at once apparent [S-K] that if there are only a finite number of measured integrals of f, whether over lines or strips, that f cannot be uniquely determined. Any procedure yielding an image $\hat{f}$, which is to be displayed as an estimate of f, given the measurements, thus must choose one among many solutions, unless there is a criterion for optimality. What is the real criterion? What properties should $\hat{f}$ have? It is not mathematics to say that $\hat{f}$ should "give useful diagnostic information", but this is the radiologists criterion. It would seem that the more accurate $\hat{f}$ is in approximating f the more likely we are to achieve the radiologists' ill-defined, subjective criterion. It is clear from the high degree of nonuniqueness above that one must consider f to be in some restricted class. We might try smooth functions, but the human anatomy is not even very continuous looking much less smooth (e.g. the discontinuous interface between the skull and the brain). We might try an optimality criterion based on some metric distance between f and $\hat{f}$, but this seems rather arbitrary. Why not try to model human anatomy with a simple superposition of ellipses [S-L]? Why not then design the system to best reconstruct the model? One should do this honestly so that neither the measurement system or the reconstruction algorithm should actually know the model. The many design choices to be made should be compared on the model. The projection integrals of the model can be computed exactly. This simulation-type method has the advantage that a very simple model of the human anatomy (with only 11 ellipses) is very convincing to radiologists as being reasonably realistic and gives a well-defined procedure for judging algorithms and measurement design alternatives. But it is empirical rather than deductive. Mathematics still plays a role in suggesting various reconstruction algorithms and measurement designs but the criterion for system performance is nonmathematical. Of course one could arbitrarily use an L_p norm on $f-\hat{f}$ but no one has succeeded in choosing a good value of p. This has been occasionally used with $p=1,2,\infty$ but has not proved useful.

The ad hoc simulation technique has been very successful [S-K] in pointing the way to good designs for CT devices and for separating artifacts due to errors in data from those introduced by the algorithm itself. Nevertheless one should be aware of the fact that these techniques can never prove anything. CT is thus an experimental rather than a deductive science, which however relies upon mathematics for suggesting good algorithms. As discussed in [S-L] two classes of algorithms are suggested by different mathematical models. Those based on a formula and those based on iteration. For the most part iterative techniques are slower and less accurate and thus present no advantages for X-ray CT. However the situation may be different for PET as we shall see.

For those who are interested in the way that a line integral of the density f of a physical object can be measured in X-ray CT to arbitrary accuracy (in principle), we end this section with an explanation of this. Suppose that an X-ray tube puts out a beam of X-ray photons beamed along a line L. Suppose at a distance s along L the density f has the value $f(s)$ and that $f(s)\Delta s$, to first order, is the probability that a photon reaches $s+\Delta s$ given that it reached s. Then the probability that the photon traverses L is the product

$$P = (1-f(s_1)\Delta s_1)(1-f(s_2)\Delta s_2) \ldots (1-f(s_n)\Delta s_n)$$

where $\Delta s_k = s_{k+1} - s_k$. Since to first order, $1-f(s)\Delta s = \exp(-f(s)\Delta s)$, and since the ratio of the output to input intensities (measurable quantities) of the X-ray beam is an arbitrarily good (as the intensities $\to\infty$) approximation to the probability p, we have

$$-\log \frac{I_{out}}{I_{in}} = \int_L f \, ds \,.$$

§3. NMR CT

Reinforcement of the thesis that sooner or later ones favorite abstract mathematics will find application in the real world can be found in our experience with the X-ray CT problem. This also holds for NMR CT. Indeed, back in his same 1917 paper, Radon [R] showed also that a function $f(x,y,z)$ can be uniquely recovered from its plane integrals $P(\pi) = \int_\pi f \, dA$, where π is a plane and dA is area on π, under rather general conditions on f. To sketch a proof [S] we form the point function

$$F(Q) = \frac{1}{4\pi} \int_S P(\pi_\omega^Q) d\omega \tag{3.1}$$

which is the integral over the unit sphere about Q of $P(\pi)$ where $\pi = \pi_\omega^Q$ is the plane through Q whose unit normal vector is ω and ω runs through the unit sphere S. It is easy to see that $F(Q)$ is a linear functional on the underlying f which can be written [R,S] as

$$F(Q) = \frac{1}{2} f * \frac{1}{r}(Q) \tag{3.2}$$

the convolution of f with $1/2r$. Indeed from the Riesz representation theorem and translation invariance F must be the convolution of f with some function $g(r)$. By specializing $f(x,y,z) = \exp(-cr^2)$ for example one sees easily that $g(r) = 1/2r$. As Radon observed, (3.2) gives F as the electrostatic potential due to charge $\frac{1}{2}f$ so that from Gauss's theorem,

$$f(Q) = -\frac{1}{2\pi} \nabla^2 F(Q) \,. \tag{3.3}$$

Similar to the case of X-ray CT, the inversion (3.2)-(3.3) is far from practically realizable if one knows $P(\pi)$ for only finitely many planes π. In fact a differently appearing inversion can be obtained which is more amenable to numerical approximation. It is shown in [S], see also [L] for a still different approach, that (3.3) may be replaced by

$$f(Q) = -\frac{1}{8\pi^2} \int_S d\omega\, P''(t,\omega) \tag{3.4}$$

where $P(t,\omega) = P(\pi)$ where π is the plane at distance $t>0$ from the origin whose unit normal has direction ω, and P'' is the second derivative. One derives (3.4) from simple Fourier analysis [S] based on the fact that the Fourier transforms of f and $P(\cdot,\,\omega)$ are equal which is familiar from the X-ray CT case [S-L]. (3.4) is much more amenable to numerical analysis than (3.3) and so it is reasonable to use as in [S] a second difference to replace the second derivative in (3.4), summing over a finite number of directions $\omega \in S$. Although the choice of directions is not as simple as in the X-ray CT case, where equi-spaced sampling can be used, because no finite set of n directions gives a natural decomposition of S for $n>20$, one can use equi-spaced longitudes and latitudes with Jacobian weighting to obtain a quadrature of (3.4). N. J. A. Sloane has suggested triangular grids on each face of an icosahedron with Jacobian weighting (but here the Jacobian is uniformly reasonably close to one) and this works fine.

The remarkable fact is that there is a way to measure the plane integrals of a physical density. As explained in [S] and the references there, an object placed in a magnetic field which is not homogeneous will have its hydrogen atoms rotating with a nuclear magnetic spin rate proportional to the strength of the magnetic field at each point. If the nonhomogeneous field strength is constant along planes, the spin energy at a given frequency will be proportional to the integral of the hydrogen spin density along the plane corresponding to that frequency. This technique has already produced images in the laboratory which are comparable to (but not yet as good or useful as) those produced by X-ray CT. But it is clearly going to be a success and will give new, orthogonal diagnostic information.

§4. PET

Suppose $f(x,y)$ (or $f(x,y,z)$) is a probability density in the plane or in space. Choose a point Q from f at random and a line L through Q whose direction is uniformly distributed over 2π radians (or 4π steradians). Suppose this is repeated independently n times and lines $L_1,\ldots,L_n$ are obtained, the points $Q_1,\ldots,Q_n$ are assumed not observable. The problem is to estimate f given the data $L_1,\ldots,L_n$.

This problem arises in position emission tomography which began over 20 years ago but still has not achieved its promise and thus remains an area of active research [B], [P], [T]. Whereas X-ray CT does not allow metabolic studies, PET can be used to learn how the body processes chemical compounds as follows. A compound such as glucose is made using carbon isotope C_{11} atoms which emit positrons. An emitted positron annihilates with a nearby electron and the result is two photons moving along a random line L through the point Q of annihilation. The two photons move at the speed of light in opposite directions along L and are detected by a bank of detectors which can record coincidences. If the detector bank is a planar ring the problem is the two-dimensional one. It is of course much more expensive to build a cylinder or sphere about the body to catch lines L in more general position but such designs and corresponding reconstruction algorithm are now under serious study.

Up to now PET reconstruction algorithms have not differed essentially from those of X-ray CT. Each pair of detector elements defines a detector tube $T = T_k$ and the number $n_{\cdot k}$ of coincidences in the kth tube is recorded and used as an estimate of the integral of f over the tube $T_k, k=1...,K$. Y. Vardi and I have recently given a new approach based on the $E-M$ procedure of mathematical statistics [D]. Let the object field be divided into boxes B_i and let N_{ik} denote the number of coincidences in tube T_k due to annihilations in box B_i. We want to estimate $\lambda_i = \int_{B_i} f$, $i=1,...,I$. Assuming that the emissions occur in accord with a Poisson process with variable rate f we can attempt to maximize $p(n|\lambda)$, the probability of seeing $n_{\cdot k} = \sum_i n_{ik}$ given λ. We have proven that $p(n|\lambda)$ has a unique maximum value and the $E-M$ procedure gives an iterative algorithm for which the likelihood $p(n|\lambda^{new}) \geq p(n|\lambda^{old})$ increases at each step. It seems that the procedure has advantages over the usual X-ray CT algorithm used for PET in less sensitivity to noise, although it is a much slower algorithm, and is hard to implement. The exact form of the algorithm which we will discuss in detail elsewhere is:

$$\lambda_i^{new} = \lambda_i^{old} \sum_{k=1}^{K} \frac{n_{\cdot k} C_{ik}}{\sum_{j=1}^{I} \lambda_j^{old} C_{jk}}$$

where $C_{ik} = P$ (an annihilation in B_i is detected in T_k). We see that the algorithm is nonlinear and iterative. One must choose an initial guess $\lambda_i^{(0)}$ for λ_i. It is nice that if $\lambda_i^{(0)} \geq 0$ then $\lambda_i^{(k)}$ remains nonnegative. We can show that $\lambda_i^{(k)}$ converges as $k \to \infty$ to a value which has the maximum possible likelihood.

References

1. [R], Radon, J. (1917) On the determination of functions from their integrals along certain manifolds (German) Ber. Saech. akad. Wiss. 69, 262-277.

2. [S-K], Shepp, L. A., Kruskal J. B., (1979) Computerized tomography: The new medical X-ray technology Am. Math. Monthly 85, 420-439.

3. [S-L], Shepp, L. A., Logan B. F., (1974) The Fourier reconstruction of a head section IEEE Trans. Nucl. Sci. NS-21, 21-43.

4. [S], Shepp L. A., (1980) Computerized tomography and Nuclear Magnetic Resonance, Jour. Comp. Asst. Tomography, 94-107.

5. [L], Lanterbur P. C. (1974) Magnetic resonance zeugmatography Pure Appl. Chem. (40) 149-157.

6. [B], Derenzo S. E., Zaklad, H., and Budinger T. F. (1975), Positron emission camera for transverse section tomography, Available from National Technical Info. Service, U.S. Department of Commerce, 5285 Port Royal Rond, Springfield, Va. 22161, $6.

7. [P], Phelps M. E., Kuhl D. E., Mazziotta J. C. (1981) Science 211, 1445.

8. [T], Ter-Pogossian M. M., Raichle M. E., Sobel B. E., Positron emission tomography, (1981) Scientific American.

9. [D], Dempster A. P., Laird N. M., Rubin D. B., (1977) Maximum likelihood from incomplete data via the EM algorithm, J Roy Stat. Soc. 39, 1,-38.

BELL LABORATORIES
MURRAY HILL, NEW JERSEY 07974

Proceedings of Symposia in Applied Mathematics
Volume 27, 1982

RECONSTRUCTION FORMULAS IN COMPUTED TOMOGRAPHY

by

Kennan T. Smith

Since their introduction in the famous papers of Ram-Lak [11] and Lak [7], explicit reconstruction formulas in Computed Tomography have become more clear and more general, and the derivations have become simple. The purpose of this article is to present explicit formulas for arbitrary translation invariant procedures. The formulas allow a free choice of the point spread function on the basis of special characteristics of the problem (e.g. noise, regions of high or low density contrasts, etc.), and then provide the resulting reconstruction kernels.

The Table of Contents is as follows:

INTRODUCTION

Two kinds of x-ray beam are used in Computed Tomography - a parallel beam, in which the x-ray photons travel along parallel lines with a given direction θ, and a divergent beam, in which the x-ray photons travel along rays from a given source point a. The data measured in an x-ray procedure are the integrals of the density function (more precisely the x-ray attenuation coefficient) along the corresponding lines or rays.

The parallel beam x-ray transform or radiograph of a function f from the direction θ at the point x in the plane $\theta^{\perp}$ perpendicular to θ is the function

$$(1.1)\qquad \mathscr{P}_{\theta} f(x) = \int_{-\infty}^{\infty} f(x+t\theta)\, dt, \qquad x \in \theta^{\perp} .$$

The divergent beam x-ray transform or radiograph of the function f from the source point a in the direction θ is the function

$$(1.2)\qquad \mathscr{D}_{a} f(\theta) = \int_{0}^{\infty} f(a+t\theta)\, dt .$$

The basic problem of Computed Tomography is to reconstruct a good approximation to the density function f from a finite (preferably not too large) number of radiographs.

This research is supported by the National Science Foundation under Grant No. MCS-8101586.

0160-7634/82/0000-0874/$05.00

In practice the 2 dimensional space R^2 and the 3 dimensional space R^3 are both relevant. To avoid duplications the discussion will be placed in the n dimensional space R^n. A direction in R^n is a point on the unit sphere S^{n-1}, i.e. a point of absolute value 1. The density function f is assumed throughout to be square integrable and to be 0 outside a bounded open set Ω.

In addition to the x-ray transforms defined above, two related transforms are needed. The line integral transform of f is the function

$$(1.3) \qquad \mathscr{L}_x f(y) = \int_{-\infty}^{\infty} f(x+t(y/|y|)) \, dt .$$

The Radon transform of f is the function

$$(1.4) \qquad \mathscr{R}_\theta f(t) = \int_{\langle x,\theta\rangle = t} f(x) \, dx , \qquad t \text{ real.}$$

The following relations are clear.

$$(1.5) \qquad \mathscr{L}_x f(y) = \mathscr{D}_x f(y/|y|) + \mathscr{D}_x f(-y/|y|)$$

$$(1.6) \qquad \mathscr{L}_x f(\theta) = \mathscr{P}_\theta f(E_\theta x) ,$$

where E_θ is the orthogonal projection (in R^n) on the subspace $\theta^\perp$.

The Fourier transform of f is given by

$$(1.7) \qquad \hat{f}(\xi) = (2\pi)^{-n/2} \int e^{-i\langle x,\xi\rangle} f(x) \, dx .$$

Under the Fourier transform, the convolution of two functions,

$$(1.8) \qquad f*g(x) = \int f(x-y) g(y) \, dy ,$$

is sent to the product of the transforms,

$$(1.9) \qquad (f*g)^\wedge(\xi) = (2\pi)^{n/2} \hat{f}(\xi)\hat{g}(\xi) .$$

The constant

$$(1.10) \qquad \gamma_n = \frac{\Gamma((n+1)/2)}{2(n-1)\pi^{(n+1)/2}}$$

recurs throughout the article.

CONVOLUTION FORMULAS

For a fixed source x the integral of $\mathscr{D}_x f$ over the sphere S^{n-1} is given by

$$\int_{S^{n-1}} \mathscr{D}_x f(\theta)\,d\theta = \int_{S^{n-1}} \int_0^\infty f(x+t\theta)\,dt\,d\theta = \int_{R^n} f(x+y)\,|y|^{1-n}\,dy \;.$$

Thus

$$(2.1) \qquad \int_{S^{n-1}} \mathscr{D}_x f(\theta)\; d\theta = R^* f(x) \;, \quad \text{with} \quad R(x) = |x|^{1-n} \;.$$

For use in potential theory, M. Riesz [12] introduced the functions

$$(2.2) \qquad R_\alpha(x) = \frac{\Gamma((n-1)/2)}{2^\alpha \pi^{n/2} \Gamma(\alpha/2)} |x|^{\alpha-n} \;, \quad 0 < \alpha < n \;.$$

The present function R is a multiple of R_1. Since R_α is positively homogenous of degree $\alpha-n$, the Fourier transform $\hat{R}_\alpha$ is positively homogenous of degree $-\alpha$, and since R_α is rotation invariant, $\hat{R}_\alpha$ is also. From these two facts (and the fact that $\hat{R}_\alpha$ is a function) it follows that $\hat{R}_\alpha(\xi)$ is a constant multiple of $|\xi|^{-\alpha}$. The mysterious constant in (2.2) is chosen so that

$$(2.3) \qquad \hat{R}_\alpha(\xi) = (2\pi)^{-n/2}\, |\xi|^{-\alpha} \;,$$

hence so that

$$(2.4) \qquad (R_\alpha * f)^\wedge\, (\xi) = |\xi|^{-\alpha}\, \hat{f}(\xi) \;.$$

Therefore, if the operator Λ is defined by

$$(2.5) \qquad (\Lambda f)^\wedge\, (\xi) = |\xi|\, \hat{f}(\xi) \;,$$

then Λ^α inverts convolution by R_α, and, for $\alpha = 1$, formula (2.1) becomes

$$(2.6) \qquad f(x) = 2\gamma_n\, \Lambda \int_{S^{n-1}} \mathscr{D}_x f(\theta)\; d\theta \;.$$

This preliminary formula is unsatisfactory in various respects, one of them being that application of the operator Λ to a function g requires the knowledge of g at all points of R^n. Consequently, (2.6) requires the x-ray data from sources throughout R^n. This particular defect is remedied by writing (2.6) a second time with θ replaced by $-\theta$, adding the result to (2.6), and using (1.5). The result is

$$(2.7) \qquad f(x) = \gamma_n\, \Lambda \int_{S^{n-1}} \mathscr{L}_x f(\theta)\; d\theta \;.$$

As opposed to $\mathscr{D}_x f$, $\mathscr{L}_x f$ is determined for all x by its values at points x on any sphere (or other $(n-1)$ manifold) surrounding Ω. For such x, $\mathscr{L}_x f$ is determined by $\mathscr{D}_x f$ in accordance with (1.5). Thus (2.7) requires x-ray data only from sources on a manifold surrounding Ω.

When combined with (1.6), formula (2.7) also gives a reconstruction formula for parallel beam x-rays:

$$f(x) = \gamma_n \Lambda \int_{S^{n-1}} \mathscr{P}_\theta f(E_\theta x)\, d\theta . \tag{2.8}$$

Remark 2.9 In the case of the Radon transform, the operator Λ^{n-1} and the Riesz kernel R_{n-1} play the roles of Λ and R_1. In the case of integrals over k planes, Λ^k and R^k play these same roles. See [14, 15]. In particular, the Radon inversion formula [10] is

$$f(x) = (1/2)(2\pi)^{1-n} \Lambda^{n-1} \int_{S^{n-1}} \mathscr{R}_\theta f(\langle x,\theta\rangle)\, d\theta . \tag{2.10}$$

Often the x-ray inversion formulas are derived from the Radon formula through changes of variable, but this works only in dimension 2, and moreover the direct derivations are much simpler. They appeared first in [5].

In the divergent beam reconstruction formula (2.7) the operator Λ which involves both differentiation and singular integration (see Appendix) is applied to the data as functions of the x-ray sources which ought to be held to a minimum. The roles of sources and readings can be reversed by using the relation

$$\mathscr{L}_x f((a-x)/|a-x|) = \mathscr{L}_a f((x-a)/|x-a|) \tag{2.11}$$

in which both sides are the integral of f along the line through x and a.

Let the actual sources "a" lie on the sphere S of radius r which surrounds Ω, let x be a fixed point either inside or outside S, and in (2.7) make the change of variable

$$\theta = (a-x)/|a-x|, \quad \text{hence} \quad d\theta = (1/r)|\langle x-a,a\rangle|\,|x-a|^{-n}\, da . \tag{2.12}$$

When the symmetry (2.11) and the homogeneity

$$\mathscr{L}_a f(ty) = \mathscr{L}_a f(y)$$

are used, formula (2.7) becomes

$$f(x) = (\gamma_n/r) \Lambda \int_S \mathscr{L}_a f(x-a)\, |\langle x-a,a\rangle|\,|x-a|^{-n}\, da. \tag{2.13}$$

In terms of the operator

$$(2.14)\qquad \tilde{\mathscr{L}}_x f(y) = \mathscr{L}_x f(y)\ |\langle y,x\rangle|\,|y|^{-n} ,$$

which also is obtainable directly from the x-ray data, formula (2.13) takes the form

$$(2.15)\qquad f(x) = (\gamma_n/r)\ \Lambda \int_S \tilde{\mathscr{L}}_a f(x-a)\ da .$$

Remark 2.16 The change of variable (2.12) is somewhat peculiar. Both for points x inside S and for points x outside S it leads to (2.13), but for quite different reasons in the two cases. More details are given in [8]. In (2.13,2.15) the singular operator Λ is applied to the x-ray data as functions of the readings, which are more or less unrestricted, rather than as functions of the sources, which should be held to a minimum.

FOURIER TRANSFORM FORMULAS

Because of translation invariance, the Fourier transform has played a preponderant role in the study of the parallel beam x-ray transform. If ρ is a function of 1 variable, then

$$(3.1)\qquad \int_{R^n} f(x)\rho(\langle x,\xi\rangle)\ dx = \int_{\theta^\perp} \mathscr{P}_\theta f(x)\rho(\langle x,\xi\rangle)\ dx\ ,\quad \xi\varepsilon\theta^\perp ,$$

as can be seen simply by writing out $\mathscr{P}_\theta f$. For $\rho(t) = e^{-it}$ this gives

$$(3.2)\qquad \hat{f}(\xi) = (2\pi)^{-1/2}\ (\mathscr{P}_\theta f)^\wedge\ (\xi),\quad \xi\varepsilon\theta^\perp .$$

In Computed Tomography this is known as the "Central Slice Theorem", but it has been known and used extensively in other disciplines since the beginnings of the Fourier transform.

Remark 3.3 With $\rho(t) = t^m$, formula (3.1) gives the consistency conditions that effectively characterize the range of the parallel beam x-ray transform. See [6,9,15].

The corresponding formula for the divergent beam is

$$(3.4)\qquad \hat{f}(\xi) = (1/2r)(2\pi)^{-n/2} \int_S e^{-i\langle a,\xi\rangle}\ \tilde{\mathscr{L}}_a f(\theta)\ da,\quad \xi\varepsilon\theta^\perp .$$

where S is a sphere of radius r surrounding Ω . To verify this, fix θ and let S_+ be the hemisphere $\langle\theta,a\rangle \geq 0$. In the integral on the right side of (3.4) let x' be the orthogonal projection of a on $\theta^\perp$, in which case $dx' = |\langle\theta,a\rangle/r|da$. The integral over S becomes

$$r\int_{\theta^\perp} e^{-i<x',\xi>} \mathscr{L}_{x'}f(\theta) = r\int_{\theta^\perp} e^{-i<x',\xi>} \mathscr{P}_\theta f(x')\, dx' .$$

The same is true in the opposite hemisphere, and combination of the two gives

$$(3.5) \qquad (\mathscr{P}_\theta f)^\wedge (\xi) = (1/2r)(2\pi)^{(1-n)/2} \int_S e^{-i<a,\xi>} \tilde{\mathscr{L}}_a f(\theta)\, da, \quad \xi\varepsilon\theta^\perp .$$

Along with (3.2), this gives (3.4).

Remark 3.6 For dimension 2, formulas (3.4,3.5) were found by the author through computation of the Fourier transforms of both sides of (2.13). The extension to dimension n and the simple proof above were found by D.V. Finch.

Formula (3.2) provides the basis of the "Fourier Reconstruction Method". If $\theta_1,\dots,\theta_m$ are the parallel beam x-ray directions, then the x-ray data provide the values $\hat{f}(\xi)$ for ξ in the subspaces $\theta_j^\perp$. The additional values needed for a discrete inverse Fourier transform are obtained by interpolation. Formula (3.4) provides a similar possibility in the case of the divergent beam transform. Alternatively, formula (3.5) can be used along with the 1 dimensional Fourier transform, a parallel beam reconstruction kernel, and back projection.

APPROXIMATE FORMULAS

Because of the finite data, exact reconstructions can never be achieved, even with exact formulas like (2.6) and (2.7). In the present situation this is even more true as the operator Λ involves both differentiation and singular integration applied to noisy data. Indeed, it is fruitful to adopt from the beginning the point of view that it is not a reconstruction of the density function f that is sought, but rather a reconstruction of an approximation $\mathscr{A}f$ to f.

If the approximation operator $\mathscr{A}$ is translation invariant, i.e. if the same reconstruction results in Cincinnati and San Francisco, then $\mathscr{A}$ has the form $\mathscr{A}f = e*f$, and if $\mathscr{A}f$ is to be an approximation to f, then e is an approximate δ-function, called in this case the point spread function. The point spread function is the result of applying the reconstruction procedure in the case where the density function f is a point mass.

Convolution of both sides of (2.15) with the point spread function e gives

$$e*f(x) = (\gamma_n/r) \int_S (\Lambda e*\tilde{\mathscr{L}}_a f)(x-a)\, da.$$

Since

$$\tilde{\mathscr{L}}_a f(ty) = |t|^{1-n} \tilde{\mathscr{L}}_a f(y)$$

it follows that

$$(\Lambda e * \tilde{\mathscr{L}}_a f)(y) = \int_{S^{n-1}} \tilde{\mathscr{L}}_a f(\theta) \int_0^\infty \Lambda e(y-t\theta) \ dt \ d\theta ,$$

and hence that

(4.1) $$e*f(x) = \int_S \int_{S^{n-1}} \tilde{\mathscr{L}}_a f(\theta) k(x-a,\theta) \ d\theta \ da , \quad \text{with}$$

(4.2) $$k(y,\theta) = (\gamma_n/r) \int_0^\infty \Lambda e(y-t\theta) \ dt = (\gamma_n/r) \Lambda \mathscr{D}_y e(-\theta) .$$

To obtain a corresponding formula for the parallel beam, replace f in (2.6) by $e*f$ and use the fact that

$$\mathscr{P}_\theta(e*f) = \mathscr{P}_\theta e * \mathscr{P}_\theta f$$

to get

(4.3) $$e*f(x) = \int_{S^{n-1}} \int_{\theta^\perp} \mathscr{P}_\theta f(y) \ k(E_\theta x-y) \ dy \ d\theta, \quad \text{with}$$

(4.4) $$k(y) = \gamma_n \ \Lambda \ \mathscr{P}_\theta e(y) .$$

The inner integral in (4.3) is the convolution of the kernel k with the data $\mathscr{P}_\theta f$. This fact has given the name "Convolution Reconstruction Method" to the formulas (4.1-4.4), although the inner integral in (4.1) only resembles a convolution. In both cases the inner integral is termed the "convolution" step, and the outer integral is termed the "back projection" step.

The standard method for producing an approximate δ-function is to choose a function E, set

(4.5) $$e_r(x) = r^{-n} E(x/r) ,$$

and take $e = e_r$ for a suitably small r. According to (1.9),

(4.6) $$(e_r*f)^\wedge(\xi) = (2\pi)^{n/2} \ \hat{E}(r\xi)\hat{f}(\xi) .$$

Therefore, if $\hat{E}$ is bounded and continuous with

$$\hat{E}(0) = (2\pi)^{-n/2} \qquad \text{then}$$

$$(4.7) \qquad e_r * f \to f \quad \text{in } L^2, \quad \text{as } r \to 0 .$$

The convergence is not uniform on the unit ball in $L^2(\Omega)$, but since the operators $f \to e_r * f$ are uniformly bounded, the convergence is uniform on any compact set (e.g. the unit ball in any Sobolev space $\mathscr{H}^s(\Omega)$ with $s > 0$). More generally, if $0 < t < s$, there is a constant $c_r = c_r(s,t)$ such that

$$(4.8) \qquad ||e_r*f-f||_{\mathscr{H}^t} \leq c_r ||f||_{\mathscr{H}^s} \quad \text{and} \quad c_r \to 0 \quad \text{as} \quad r \to 0 ,$$

the norms being those in the spaces $\mathscr{H}^t(\Omega)$ and $\mathscr{H}^s(\Omega)$. (The Sobolev spaces are discussed in detail in [1].)

If E is bounded and integrable with integral 1, and f is bounded, then

$$(4.9) \qquad e_r * f \to f \quad \text{almost everywhere on } \Omega ,$$

including every point of continuity, and the $e_r * f$ are dominated on Ω by an L^2 function independent of r. If E has compact support, the same is true whether f is bounded or not. Statement (4.9) is a consequence of the Lebesgue theorems on differentiation, and the uniform domination is provided by the Hardy-Littlewood maximal theorem. (See [13]).

Subject to the above limitations (basically, integral 1, or $\hat{E}(0) = (2\pi)^{-n/2}$), the function E and the number r can be chosen to suit particular features of the problem at hand and the kernel k can be computed with formulas (4.2, 4.4). In general the kernel depends on θ in (4.4) as well as in (4.2). Ordinarily, but not necessarily, the point spread function is independent of θ, in which case the parallel beam kernel is independent of θ, and the divergent beam dependence can be removed through the formula

$$\mathscr{D}_{Ua} e(U\theta) = \mathscr{D}_a e(\theta)$$

where U is any orthogonal transformation. It is not mandatory to use the same point spread function everywhere. Within regions of established high contrast it might conceivably be useful to use point spread functions supported by cones which avoid averaging over the interfaces.

In computed tomography a different route has been followed. Formula (4.4) is equivalent to

$$(4.10) \qquad \hat{k}(\xi) = (2\pi)^{1/2} \gamma_n \hat{e}(\xi) |\xi| .$$

Thus $\hat{k}$ is the product of $|\xi|$ with a "cutoff function" $\hat{e}$. In computed tomography e has been chosen primarily on the basis of the apparent suitability of $\hat{e}$ as a cutoff function.

Remark 4.11 The original convolution formula, found by Ram-Lak [11], uses the cutoff function $\hat{e}(\xi) = 1$ for $|\xi| < R$, 0 for $|\xi| > R$, which leads to

$$(4.12) \qquad E(x) = (1/2|x|)\, J_1\,(|x|).$$

The general formulas (4.1-4.4) were developed during a series of lectures at the Mayo Clinic under a grant from the Waters Instrument Co. Two dimensional cases were found subsequently, but independently, by C. Lantejoul. Whether the original divergent beam formula of Lak [7] is a particular instance of (4.1, 4.2) has not been investigated.

Remark 4.13 The derivation of (4.3, 4.4) also gives the approximate inversion formulas for the Radon transform:

$$(4.14) \qquad e*f(x) = \int_{S^{n-1}} \int_{-\infty}^{\infty} \mathscr{R}_\theta f(t)\, k(\langle x,\theta\rangle - t)\, dt d\theta ,$$

$$(4.15) \qquad k(t) = (1/2)(2\pi)^{1-n}\, \Lambda^{n-1}\, \mathscr{R}_\theta e(t) .$$

RESTRICTED X-RAY SOURCES AND DIRECTIONS

The approximate reconstruction formulas of the last section involve integrals over a full sphere of x-ray sources or directions, but some applications allow only sources or directions in some fixed portion of the sphere (usually an arc). At the present time, although there are iterative procedures to handle this situation, there are no explicit formulas, either exact or approximate. However, there is elegant approximate inversion formula for the Radon transform due to Davison and Grunbaum [4]. In dimension 2 the Radon transform is equivalent to the parallel beam x-ray transform, so the Davison-Grunbaum procedure applies to the parallel beam x-ray transform in dimension 2. By virtue of formula (3.5), it applies also to the divergent beam x-ray transform in dimension 2 in cases where a sufficiently wide arc of sources is available.

If $\theta_1,\ldots,\theta_m$ are the directions in which the Radon transform is measured and e is the desired point spread function, the Davison-Grunbaum approximate solution operator $\mathscr{S}$ is defined by

$$(5.1) \qquad \mathscr{S}f(x) = \sum_{j=1}^{m} \mathscr{R}_{\theta_j} f * k_j(\langle x,\theta_j\rangle) \qquad \text{with}$$

$$k = (k_1,\ldots,k_m) \qquad \text{satisfying}$$

(5.2) $\mathscr{R}\mathscr{R}^*k = \mathscr{R}e = (\mathscr{R}_{\theta_1}e, \ldots, \mathscr{R}_{\theta_m}e)$.

It is easily seen that $\mathscr{S}f = f*\mathscr{R}^*k$ so that $e*f-\mathscr{S}f = f*(e-\mathscr{R}^*k)$ and hence that

(5.3) $\max |e*f-\mathscr{S}f| \leq ||f||_{L^2} \; ||e-\mathscr{R}^*k||_{L^2}$.

The essential feature of this procedure is that when Ω is a disk in the plane the kernels k and the error bound (5.3) can be computed explicitly.

Remark 5.4 The condition (5.2) means simply that k is chosen so that $\mathscr{R}^*k$ is the orthogonal projection of e on the range of $\mathscr{R}^*$. The Davison-Grunbaum procedure is one illustration of the value of looking for an approximation to an approximate solution rather than an approximation to an exact solution. In the case of an exact solution the point spread function e would be a δ-function which could not be projected on the range of $\mathscr{R}^*$. The idea of looking for an approximation to an approximate solution rather than for exact formulas may be an important one. From the point of view of the author it originated in the article of Davison-Grunbaum, but they credit their perception of it to the article of Cho [3].

EXAMPLES

The work in this section was done jointly with M. Bottema and D.C. Solmon and with indispensable help from A. Coppola and R. Eifrig. The pictures were made possible by the generosity of the Tektronix Co. in providing a high resolution monitor with 16 gray levels and a camera for hard copy. Initially it was intended to do both parallel and divergent beam reconstructions, the latter via parallel beam kernels and formula (3.5). However, time ran out.

As mentioned in section 2, the standard method for producing an approximate δ-function e is to choose a function E with integral 1 or with $\hat{E}(0) = (2\pi)^{-n/2}$, set

$$e_r(x) = r^{-n}E(x/r) ,$$

and take $e = e_r$ for a suitably small r. If K is the kernel corresponding to E, then $k = k_r$ is the kernel corresponding to e. In computed tomography the choices are usually made on the basis of the apparent suitability of $\hat{e}$ as a cutoff function of the Fourier side, or, more accurately, on the basis of the apparent suitability of $\hat{e}(\xi)|\xi|$ as an approximation to $|\xi|$ for small $|\xi|$. (See (4.10).)

In the original convolution formula of Ram-Lak [11]

(6.1) $\hat{E}(\xi) = 1$ for $|\xi| < 1$, 0 for $|\xi| > 1$, so that
$E(x) = (1/2|x|)J_1(|x|))$.

For $y > 0$, the corresponding (even) kernel is given by

(6.2) $$K(y) = (1/4\pi^2)((1/y)\sin y - (2/y^2)\sin^2(y/2))$$

The Shepp-Logan point spread function [13] is determined by

(6.3) $$\hat{E}(\xi) = (1/2\pi)|(2/|\xi|)\sin(|\xi|/2)|^3$$

and the corresponding (even) kernel is determined by

(6.4) $$\hat{K}(\xi) = (2\pi)^{-3/2}|\sin(\xi/2)|((2/\xi)\sin(\xi/2))^2, \quad \text{for } \xi > 0 .$$

According to [13] this produces the kernel

(6.5) $$K(j) = -1/\pi^2(4j^2-1) \quad \text{for } j \text{ an integer}$$

with K linear on the intervals between integers. According to (4.4) the point spread function is given by

(6.6) $$E(x) = \int_{S^1} K(E_\theta x)\, d\theta = 4\int_0^{|x|} K(t)(|x|^2-t^2)^{-1/2}\, dt.$$

Choosing $\hat{E}$ instead of E usually leads to a point spread function with unbounded support. The Ram-Lak point spread function, for example is not integrable. (This is the reason for allowing the condition $\hat{E}(0) = (2\pi)^{-n/2}$ in place of integral $E = 1$.) The Shepp-Logan point spread function is integrable, but since the third derivatives of $\hat{E}$ are discontinuous, it follows that $|x|^3E$ is not integrable. (The exact behavior at infinity seems to be tricky to find.) A natural family of point spread functions with bounded support is given by

(6.7) $$E(x) = \frac{\pi^{-n/2}\mu\Gamma(\mu/2)}{(\mu-n)\Gamma((\mu-n)/2)}(1-|x|^2)^{(\mu-n)/2}$$
$$E(x) = 0 \quad \text{for } |x| > 1 .$$

It is shown in [8] that for $n = 2$, $\mu = 2m + 3$, and $y > 0$ the corresponding (even) kernel is

(6.8) $$K(y) = \mu\Gamma(\mu/2)/4\pi^{5/2}m!\ [H(y)\log(|y+1|/|y-1|) + \sum_{i=1}^{\mu-2} \frac{(-1)^i}{i\, i!} H^i(y)((y+1)^i - (y-1)^i)] ,$$

with $H(y) = y(1-y^2)^m$. These will be called "logarithmic kernels." While the expression (6.8) is not very promising, the kernels have simple and very quickly converging series:

$$K(y) = \begin{cases} \dfrac{(-1)^{m+1} 1\cdot 3\cdots\mu}{4\pi^2} \displaystyle\sum_{j=0}^{\infty} y^{2j}/(2j-1)(2j-3)\ldots(2j-2m-1) \\ \dfrac{-1\cdot 3\cdots\mu}{4\pi^2} \displaystyle\sum_{j=1}^{\infty} y^{-2j}/(2j+1)(2j+3)\cdots(2j+2m+1) \end{cases}$$

according as $y < 1$ or $y > 1$.

Whatever the chosen kernel k, the first step in a reconstruction is the numerical evaluation of $\mathcal{P}_\theta f*k$ in the approximate inversion formula (4.3). This takes the form of a discrete convolution

(6.9) $$\bar{g}*\bar{k}(\ell) = \sum_j \bar{g}(j)\bar{k}(\ell-j)$$

where $\bar{g}(j)$ is determined by the values of $\mathcal{P}_\theta f$ near the point jh, $j = 0,\pm 1,\pm 2, \ldots$, and $\bar{k}$ is a sequence associated with the chosen kernel k, while $\bar{g}*\bar{k}(\ell)$ is the numerical approximation to $\mathcal{P}_\theta f*k(\ell h)$. Sometimes $\bar{k}(j) = k(jh)$, but a necessary condition on the sequence $\bar{k}$ is that

(6.10) $$\sum_j \bar{k}(j) = 0.$$

In the Ram-Lak case $\hat{k}$ vanishes for $|\xi| > r$, and expansion in a Fourier series gives

$$\hat{k}(0) = (1/2R) \sum_j k(j\pi/R), \quad R > r .$$

Since $\hat{k}(0) = 0$, (6.10) is satisfied with $\bar{k}(j) = k(jh)$ whenever the step h is less than π/r. In the Shepp-Logan case, with $k = k_r$ and $\bar{k}(j) = k(jh)$, (6.10) is satisfied if r is an integer multiple of h. In the general case (6.10) is not satisfied with $\bar{k}(j) = k(jh)$.

A general kernel k can be handled by requiring that the discrete convolution (6.9) give the exact value of the analytic convolution $g*k(\ell h)$ whenever g is a special type of function (e.g. piecewise constant or piecewise linear) taking the value $\bar{g}(j)$ at the point jh. For example, if g takes the value $\bar{g}(j)$ on the interval from $(j-1/2)h$ to $(j+1/2)h$, then

(6.11) $$\bar{g}*\bar{k}(\ell h) = g*k(\ell h)$$

with

$$(6.12) \qquad \bar{k}(j) = \int_{(j-1/2)h}^{(j+1/2)h} k(y)\,dy$$

Similarly, if g is piecewise linear, then (6.11) holds with

$$(6.13) \qquad \bar{k} = \bar{k}_1 + (\tau - 1)(\bar{k}_2 - \bar{k}_3)$$

where

$$(6.14) \qquad \bar{k}_1(j) = \int_{(j-1)h}^{jh} k(y)\,dy, \qquad \bar{k}_2(j) = j\bar{k}_1(j)$$

$$\bar{k}_3(j) = (1/h)\int_{(j-1)h}^{jh} yk(y)\,dy ,$$

and τ shifts the indices 1 unit to the right. With most kernels, the integrals in (6.12, 6.14) can be computed to any desired accuracy so that (6.11) will hold to any desired accuracy. In the computations done so far (6.13, 6.14) have proved somewhat superior to (6.12). P.M. Anselone has pointed out that the latter is a special case of the "product integration" introduced for a very different purpose by K.E. Atkinson.

Shown below are a few sample reconstructions of a phantom head section similar to the one designed by Shepp and Logan [13]. The features are all elliptical. (The apparently flat sides come from the discrete representation on the screen.) The skull has axes 128 × 100, while the tiny tumor in the group of three at the upper right is a circle of diameter 1. The densities are as follows: skull, 200; calcified pineal gland to the right of the center, 150; background brain tissue, 102; ventricles, 100; hemorrhage next to skull at upper left, 101; large circular tumor at upper left, 104; overlap of tumor and ventrical, 102; largest of three tumors at upper right, 103; other two tumors at upper right, 104. All three of the reconstructions were made with 90 parallel beam x-rays computed mathematically at 2° intervals starting at .5° from the horizontal. Each x-ray was sampled at 256 equally spaced points.

The only difficulty in reconstructing this phantom is to resolve the tiny tumor while avoiding the streaks that radiate from small regions of high density when the number of x-rays is minimal. Remnants of these streaks are clearly visible in the lower left part of the reconstruction with $\mu = 15$ and $r = 1.7$, and somewhat less so in the same part of the reconstruction with the Shepp-Logan kernel, but have pretty well disappeared with $\mu = 15$ and $r = 1.8$. A steep narrow point spread function such as

$\mu = 7$, $r = 1$, provides good resolution but lots of streaking. Graphs of the point spread functions with $\mu = 7$, $r = 1$; $\mu = 15$, $r = 1.7$; and $\mu = 15$, $r = 1.8$ are shown together below, as are graphs of the Shepp-Logan point spread function and a repetition of the one with $\mu = 15$, $r = 1.8$.

Because of the sensitivity of this particular feature of the reconstruction to small changes in the point spread function (e.g. $r = 1.7$ vs. $r = 1.8$) it seemed inadvisable to try the Ram-Lak kernel without knowing exactly what is recommended by experienced users. Explicit directions for the Shepp-Logan kernel are given in [13].

Remark 6.15 When I asked Larry Shepp for permission to include the Shepp-Logan kernel among the examples, he gave it immediately, but asked that I make clear that he had never promoted that kernel as being better than others. I want to make the same point with regard to the logarithmic kernels introduced above. The two parameters provide an easy way to change the shape of the point spread function in a rational way in order to illustrate the fact that the shape does make a difference. I feel now that they still do not provide enough flexibility.

Remark 6.16 The 3 dimensional Radon transform is the data in nuclear magnetic resonance reconstruction. (See the article of L. Shepp in this volume.) If the point spread function E is given by (6.7), then, according to (4.15) the reconstruction kernel is

$$(6.17)\quad K(t) = \frac{\mu\Gamma(\mu/2)}{2^3\pi^{5/2}\Gamma((\mu-1)/2)}(1-t^2)^{(\mu-5)/2}((\mu-2)t^2-1), |t| \le 1.$$

Point Spread Functions

$\mu=7$ $r=1$, $\mu=15$ $r=1.7,1.8$ $\mu=15$ $r=1.8$, Shepp-Logan

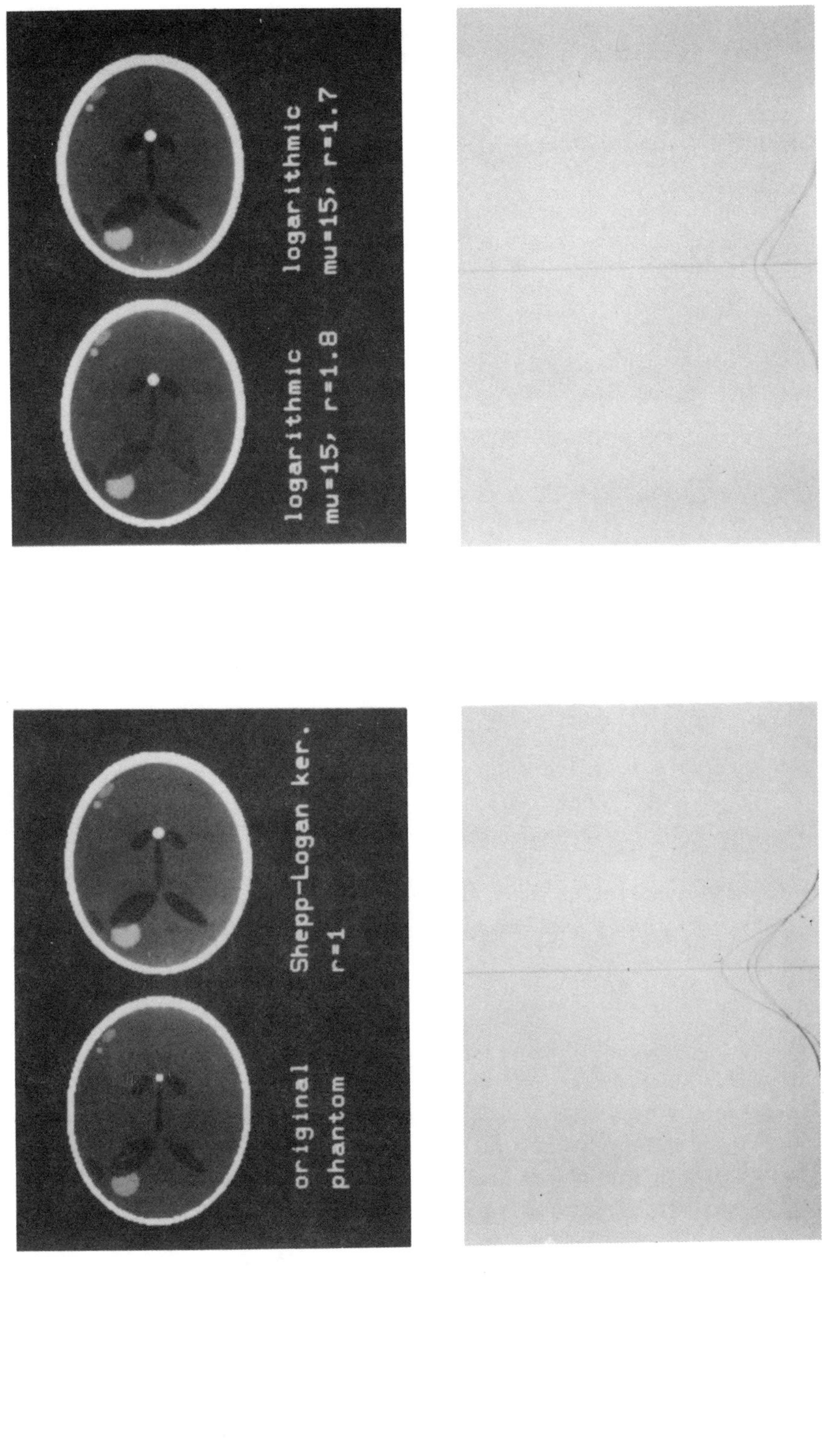

APPENDIX ON Λ

The operator Λ which appears in the reconstruction formulas is defined by

$$(7.1)\qquad (\Lambda f)^{\wedge}(\xi) = |\xi|\ \hat{f}(\xi)\ .$$

In computing the kernels for the approximate formulas it is sometimes convenient to have a direct definition that does not involve the Fourier transform.

Let

$$(7.2)\qquad H_j(x) = \frac{\Gamma((n+1)/2)}{\pi^{(n+1)/2}} x_j |x|^{-(n+1)}\ ,$$

and let $v.p.H_j$ be the distribution defined by the Cauchy principal value

$$(7.3)\qquad \langle v.p.H_j, g\rangle = \lim_{\varepsilon\to 0} \int_{|x|>\varepsilon} H_j(x) g(x)\, dx\ .$$

For $n > 1$, H_j is the pointwise derivative $-(\partial/\partial x_j)R_1$ and the distribution $v.p.H_j$ is the corresponding distribution derivative. Consequently

$$(7.4)\qquad (v.p.H_j)^{\wedge}(\xi) = -(2\pi)^{-n/2}\ i\xi_j|\xi|^{-1}\ ,$$

and this is also valid for $n = 1$. From (7.4) it follows that

$$(7.5)\qquad \Lambda f = \sum_{j=1}^{n} (\partial/\partial x_j)\ v.p.H_j * f\ ,$$

where the convolution is a distribution convolution, or, equivalently, a Cauchy principal value convolution:

$$(7.6)\qquad v.p.H_j * f(x) = \lim_{\varepsilon\to 0} \int_{|y|>\varepsilon} H_j(y) f(x-y)\, dy\ .$$

Such integrals have been studied extensively by A.P. Calderón and A. Zygmund under the name "singular integrals". The long series of articles developing the properties of these integrals and their applications to partial differential equations is one of the highlights in mathematics. In [2], the first of the series, it is shown that for f in L^p, $1 < p < \infty$, the limit in (7.6) exists both in L^p and pointwise almost everywhere.

REFERENCES

1. R. Adams, N. Aronszajn, and K.T. Smith. Theory of Bessel potentials. Part II. Ann. Inst. Fourier. 1967 1-135.
2. A.P. Calderón and A. Zygmund. On the existence of certain singular integrals. Acta Math. 1952 85-139.
3. Z.H. Cho, I. Ahn, C. Bohm, and G. Huth. Computerized image reconstruction methods with multiple photon/x-ray transmission scanning. Phys. Med. Biol. 1974 511-522.
4. M.E. Davison and F.A. Grunbaum. Convolution algorithms for arbitrary projection angles. IEEE Trans. Nucl. Sci. 1979 2670-2673.
5. C. Hamaker, K.T. Smith, D.C. Solmon, and S.L. Wagner. The divergent beam x-ray transform. Rocky Mountain J. Math. 1980 253-283.
6. S. Helgason. The Radon Transform. Progress in Mathematics No. 5. Birkhauser, Boston, 1980.
7. A.V. Lakshminarayanan. Reconstruction from divergent x-ray data. SUNY Tech. Report No. 92, Comp. Sci. Dept., Buffalo, NY. 1975.
8. J.V. Leahy, K.T. Smith, and D.C. Solmon. Uniqueness, nonuniqueness, and inversion in the x-ray and Radon problems. Internat. Symp. on Ill-Posed Problems. Univ. of Delaware, Newark, DE. 1979.
9. D. Ludwig. The Radon transform on Euclidean space. Comm. Pure and Appl. Math. 1966 49-81.
10. J. Radon. Über die Bestimmung von Funktionen durch ihre Integralwerte längs gewissen Mannigfaltigkeiten. Ber. Verh. Sachs. Akad. Wiss. Leipzig, Math.-Nat. kl. 1917 262-277.
11. N. Ramachandran and A.V. Lakshminarayanan. Three-dimensional reconstruction from radiographs and electron micrographs: application of convolutions instead of Fourier transforms. PNAS USA 1971 2236-2240.
12. M. Riesz. Intégrales de Riemann-Liouville et potentiels. Acta Szeged. 1938 1-42.
13. L.A. Shepp and B.F. Logan. The Fourier reconstruction of a head section. IEEE Trans. on Nuclear Science. 1974 21-43.
14. K.T. Smith. A generalization of an inequality of Hardy and Littlewood. Canad. J. Math. 1956 157-170.
15. K.T. Smith, D.C. Solmon, and S.L. Wagner. Practical and mathematical aspects of reconstructing objects from radiographs. BAMS 1977 1227-1270.
16. D.C. Solmon. The x-ray transform. J. Math. Anal. & Appl. 1976 61-83.

DEPARTMENT OF MATHEMATICS
OREGON STATE UNIVERSITY
CORVALLIS, OREGON 97331

Proceedings of Symposia in Applied Mathematics
Volume 27, 1982

ATTENUATED AND EXPONENTIAL RADON TRANSFORMS

Oleh J. Tretiak

ABSTRACT. Single Photon Emission Computed Tomography leads to integral equations similar to the Radon Transform. The Exponential Radon Transform applies to the relatively simple case of a uniform attenuating medium. Inversion results for this problem are reviewed and the range of the mapping is characterized. The Attenuated Radon Transform involves two functions on the plane and present both inversion and identification problems. Results pertaining to the identification problem are reviewed.

There are three kinds of computer tomography in current medical use: CAT, computer assisted tomography, which uses X-rays and had found great application; PECAT, positron emission computer tomography, which utilizes positron annihilation radiation and is a glamorous specialty since it has both great promise in medical research and is very expensive; and single photon emission tomography which is the Cinderella of the tomographic family. Single photon tomography qualifies for this role for several reasons: in spite of its primacy (it is the earliest computer tomography, see Kuhl et al), it has not acheived much eminence or practical application, as it does not even have an acronym of its own. Like Cinderella, emission tomography must work much harder than her sisters: The integral equations which govern it were not solved by Radon, and though some progress has been made during the past five years, there is still much about this problem which is not understood. In this paper we will discuss some of the mathematical results pertaining to problems of emission computed tomography.

Consider a two dimensional plane through a three dimensional body. If the attenuation coefficient in this plane is given by $\mu(x)$, $x \in R^2$, then the rays from point x_0 in direction $\omega^{\perp} = (-\sin\phi, \cos\phi)$ is attenuated by a factor $(M\mu)(x_0,\omega)$ where M is the divergent beam transform [Hamaker et al]

$$(M\mu)(x_0,\omega) = \int_0^{\infty} dt\ \mu\ (x_0 + t\omega^{\perp}) \tag{1}$$

If f(x) is the distribution of radioactivity in the plane, then the radiation outside the body, along the line $x\cdot\omega = s$, $s \in R$, $\omega = (\cos\phi,\sin\phi)$ is the

0160-7634/82/0000-0875/$03.25

contribution of all radiators along this line, and is given by

$$g(s,\omega) = (R_\mu f)(s,\omega) = \int_{L(s,\omega)} f(x)e^{-(M\mu)(x,\omega)}dt \tag{2}$$

R_μ is known as the <u>attenuated Radon Transform</u>. This leads to the following problems:

Given μ and $R_\mu f$, find f.(The inversion problem).

Given f and $R_\mu f$, find μ. (The identification problem).

Given $R_\mu f$, find μ and f. (The identification - inversion problem).

Note that the inversion problem requires the solution of a rather formidable integral equation which is linear in the unknown function. The identification problem involves us in a nonlinear integral equation.

It seems therefore tantamount to hubris to attempt the third problem: in effect, we are asked to find both the kernel of an integral equation, and to solve the equation. In spite of these seeming difficulties F. Natterer has devoted considerable energy to this problem and has obtained some interesting and remarkable results.

As far as the arbitrary inversion problem goes, several numerical algorithms have been developed and they are reviewed by Budinger et al. In these methods one samples $\mu(\cdot)$ and $f(\cdot)$ and $g(\cdot)$ an appropriate lattices of points, and approximated the Attenuated Radon Transform by a matrix. As is usual with problems of tomography, the resulting matrix is both large and sparse, so that the equations can be solved by some suitable iterative method. These algorithms, however, are substantially slower then analytically based procedures such as the convolutional algorithm [Ramachandran and Lakshminarayanan], so there is room for improvement.

If the linear attenuation coefficient is both unknown and constant, the problem becomes much easier. Assume that the object to be examined (i.e., the support set Ω for both μ and f) is bounded and convex as shown in Figure 1, and that μ is equal to a constant for $x \in \Omega$ (inside the object) and is equal to zero otherwise. Evidently

$$(M\mu)(x,\omega) = \mu(d(s,\omega) - \omega^{\perp}\cdot x)$$

where $d(s,\omega)$ is the distance between the point at which the ray leaves the oject and the line ωt (see Figure 1), and μ is the constant attenuation coefficient.

Consequently,

$$(R_\mu f)(s,\omega) = \int_{L(s,\omega)} f(x)\exp(\mu(\omega^{\perp}\cdot x - d(s,\omega))dx \tag{3}$$

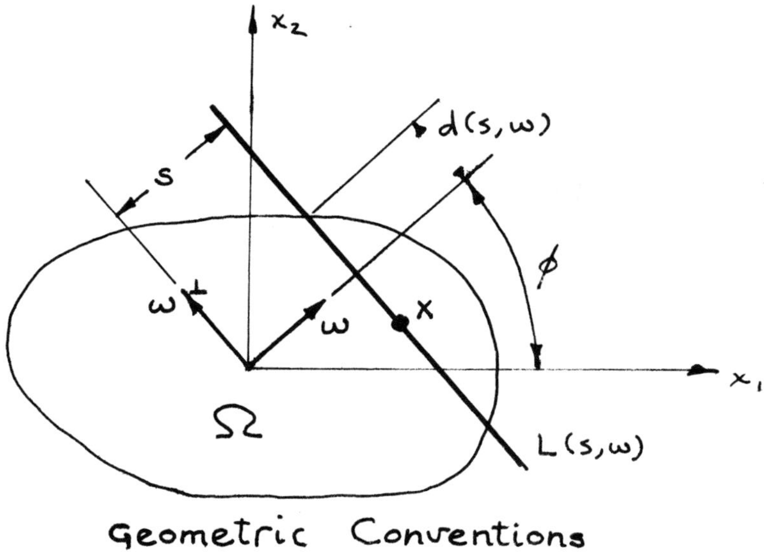

Figure 1

We introduce the shorthand

$$\int_{L(s,\omega)} dx\,h(x) = \int_{R^2} dx\ \delta(s-\omega\cdot x)\,h(x)$$

where δ is the one-dimensional Dirac measure. By evaluation we see that

$$(E_\mu f)(s,\omega) \equiv \int dx\ f(x)e^{\mu\omega^\perp\cdot x}\ \delta(s-\omega\cdot x) = e^{\mu d(s,\omega)}(R_\mu f)(s,\omega) \qquad (4)$$

Equation (4) defines the Exponential Radon Transform (ERT). We apologize for the notation since in (4) we use μ to represent both the attenuation distribution, i.e., a real valued function with support on the plane, and the real number μ which is the constant attenuation coefficient over the set Ω. In the discussion of the ERT we will use μ in the latter sense. Observe that if Ω and the scalar μ are known, one can compute $E_\mu f$ from $R_\mu f$, so that a procedure which inverts E_μ is of practical use in single photon emission tomography.

Let us reflect on the character of E_μ: it is quite similar to the Radon transform, and in fact E_0 is very much like the Radon transform. Observe, however, that the natural range of E_μ is not the same as the range of the Radon transform:

While

$$(Rf)(s,\omega) = (Rf)(-s,-\omega)$$

so that the support of the Radon transform is the set of (undirected) lines in the plane, there is no such obvious relation for the ERT, so that the support set for its range is the cylinder $R \times S^1$.

In the search for an inversion procedure for the ERT we should look for a natural mapping from functions on the cylinder to functions on the plane. This is provided by the transformation E_μ^* defined by

$$(E_\mu^* g)(x) = \int_0^{2\pi} d\phi\ g(x\cdot\omega,\omega)e^{\mu x\cdot\omega^{\perp}} \tag{5}$$

Indeed, with the appropriate Hilbert space structure on the range and domain of E_μ, we find the E_μ^* is the adjoint of E_μ. Hilbert space properties of R_μ have been explored by Gullberg.

Consider the mapping $E_{-\mu}E_\mu$. Observe the under rather simple regularity conditions of f, we obtain

$$(E_{-\mu}E_\mu f)(x) = \int_0^{2\pi} d\phi \int_{R^2} dx'\ f(x')\ \delta((x-x')\cdot\omega)\exp((x-x')\cdot\omega^{\perp}))$$

$$= \int_{R^2} dx'\ F(x')\ \frac{2\cosh \mu|x-x'|}{|x-x'|} \tag{6}$$

This equation is rather encouraging, since it is the familiar convolution integral. Unfortunately the convolution kernel is rather unattractive: it grows exponentially with $|x|$ and does not have a Fourier transform. However, there is a related result which is useful. We use the following notation for convolution on the plane

$$(f_1 \circ_2 f_2)(x) = \int_{R^2} dx'\ f_1(x-x')f_2(x')$$

and convolution on the cylinder is defined as follows:

$$(g_1 \circ_1 g_2)(s,\omega) = \int_R ds' g_1(s-s',\omega)g_2(s',\omega)$$

In terms of these definitions, one can obtain the following result

$$E_{-\mu}^* h \circ_1 E_\mu f = (E_{-\mu}^* h) \circ_2 f \tag{7}$$

This relation is expressed through the functional diagram in Figure 2. This result also suggests the following algorithm, which is very similar to the convolution algorithm used in CAT, and which we call the Exponential Convolution Algorithm (ECA).

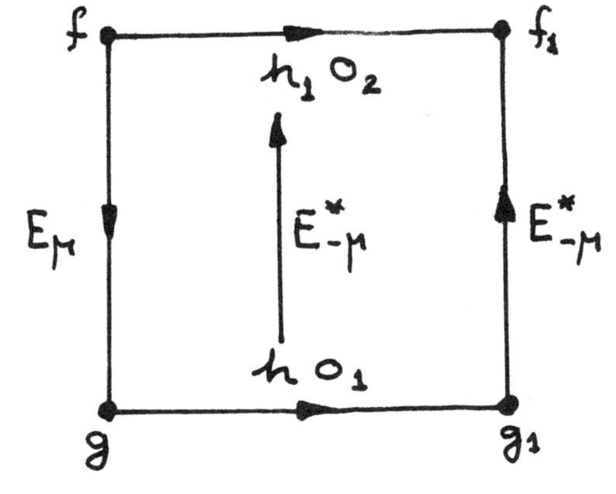

Convolutions and the ERT

Figure 2

1) Given $g_1 = R_\mu f$ where $\mu(x) = \mu\ \chi_\Omega(x)$ with μ a known real constant, and Ω a known convex bounded set. Compute
$g_2(s,\omega) = g_1(s,\omega)e^{\mu d(s,\omega)}$ where $d(s,\omega)$ is defined in Figure 1.

2) Compute
$g_3 = h\ o_1\ g_2$

3) Compute
$f_1 = E^*_{-\mu} g_3$

From the above, it follows that $f_1 = f\ o_2\ (E^*_{-\mu}h)$.

With the appropriate choice of h it might be possible to obtain $f_1 = f$. What is this desired function? Equation (7) suggests that we might look for the function which satisfies

$$(E^*_{-\mu}h)(x) = \delta_2(x)$$

where $\delta_2(x)$ is the two dimensional Dirac measure. It should not be surprising that h is a generalized function or distribution, and it can be described by Fourier Transform. An appropriate h is characterized by

$$\hat{h}(\sigma) = \begin{cases} \frac{|\sigma|}{4\pi} & |\sigma|>|\mu| \\ 0 & |\sigma|<|\mu| \end{cases}$$

The above result can be developed into an algorithm which produces satisfactory accuracy. [Tretiak and Delaney]. However, if the data $g(x,\omega)$ are noisy, the reconstruction has a surprisingly large noise variance

(Tretiak and Delaney). It is therefore worthwhile to analyze these transformation in more detail. The following results are derived by Tretiak and Metz, though Bellini et al, obtained the same formulae by a simpler method.

We define the two dimensional Fourier transform by

$$\hat{f}(u) = \int_{R^2} dx\ f(x)e^{-iu\cdot x}$$

And the n-th circular harmonic is defined by

$$\hat{f}_n(\rho) = \int_0^{2\pi} d\phi\ \hat{f}(\rho\omega)e^{-in\phi}$$

If f is bounded and of compact support f(u) is defined and analytic in all u in C^2. We similarly define the Fourier transforms on the cylinder and their circular (cylindrical) harmonics:

$$\hat{g}(\sigma,\omega) = \int_R ds\ g(s,\omega)e^{-is\sigma}$$

$$\hat{g}_n(\sigma) = \int_0^{2\pi} dy\ g(\sigma,\omega)e^{in\phi}$$

We then obtain the following result

$$(E_\mu f)\hat{}_n(\sigma) = q^n(\sigma,\mu)f_n\ (\sqrt{\sigma^2-\mu^2}) \tag{8}$$

where

$$q(\sigma,\mu) = \frac{\sigma-\mu}{\sqrt{\sigma^2-\mu^2}}$$

Observe that for $|\sigma|<|\mu|$ $E_\mu f$ depends on values of f(u) for complex u.

We see that the range of E_μ is rather restricted. Specifically, we observe that $(E_\mu f)\hat{}_n(\sigma)$ and $(E_\mu f)\hat{}_n(-\sigma)$ are related in a rather obvious way. There is a more subtle constraint on this function between the intervals $|\sigma|<|\mu|$ and $|\sigma|\geq|\mu|$. Specifically, knowledge of $(E_\mu f)\hat{}_n(\sigma)$ for $|\sigma|\geq|\mu|$ suffices for us to compute $\hat{f}_n(\rho)$ for $\rho \geq 0$, and thus to determine f(x), so that consequently we can compute $(E_\mu f)\hat{}_n(\sigma)$ for all (complex) σ.

The above results tempt us to seek for a similar representation for the mapping E_μ^*. There is a difficulty, since it is easy to see that in many circumstances $E_\mu^* g$ may not have a Fourier transform. It is, however, possible to obtain the following result: Let $g(s,\omega)$ be an absolutely integrable function with support on the cylinder, and $g_1(s,\omega)$ is related to $g(s,\omega)$ by:

$$\hat{g}_1(s,\omega) = \begin{cases} g(s,\omega) & |s|>|\mu| \\ 0 & |s|<|\mu| \end{cases}$$

In effect, we construct g_1 by eliminating all Fourier components for frequencies below $|\mu|$. Then circular harmonics of $E_\mu^* g_1$ are given by:

$$(E_\mu^* g_1)\hat{}_n(\sqrt{\sigma^2-\mu^2}) = \frac{2\pi}{\sigma}\,(q^n(\sigma,\mu)\hat{g}_n(\sigma) + q^n(-\sigma,\mu)\hat{g}_n(-\sigma)) \tag{9}$$

Note that the frequency translation is exactly opposite to that produced by E_μ (see (8)). In this context we might ask: What happens to the components of $\hat{g}(\sigma,\omega)$ for $|\sigma|<|\mu|$?E_μ^* maps these into exponentially increasing functions of $|x|$. The effect of E_μ and E_μ^* is the frequency of the Fourier Transform harmonics as shown in Figure 3.

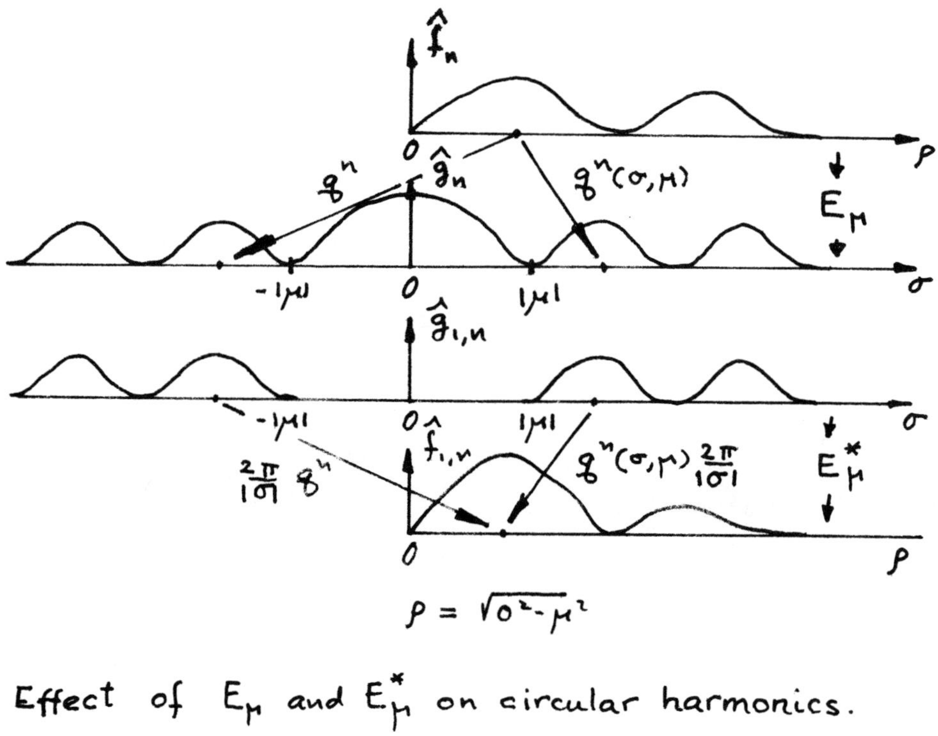

Figure 3

The above result, i.e. (9), explains the Exponential Convolution Algorithm. We can verify that

$$q(\sigma,\mu) = q^{-1}(\sigma,-\mu)$$

so that $E_{-\mu}^*$ corrects for the multiplicative factor q^n in (8), and the convolving function rejects frequencies $|\sigma|<|\mu|$ and corrects for the multiplying factor in (9).

We now turn to the identification - inversion problem. Censor et al had considerable success in computer experiments taking a purely algebraic approach. From (8) we see that (at least in this case) $R_\mu f$ is highly redundant. Natterer has devoted considerable effort to this question, and has obtained the following result:

Let $g = R_\mu f$ with $f \varepsilon C_0^\infty (R^2)$

Then, for $k > m \geq 0$

$$\int_R ds \int_0^{2\pi} d\phi \; s^m \exp(ik\phi + \frac{1}{2}(I + H)R_\mu) \, g(s,\omega) = 0 \tag{10}$$

where R_μ is the Radon transform of $\mu(x)$ and H is the Hilbert Transform. This result is related to the following interesting and remarkable result:

Let f be a linear combination of a finite number of Dirac measures (point sources). Then f can be found from g, except for an unknown constant multiplier.

Furthermore, if any of the sources are outside the attenuating region, then the multiplier is equal to one. Natterer also reports some numerical results with an algorithm based on (10). Enthusiasm for this line of research must be tempered by the following negative example: if both f and μ are circularly symmetric about the origin (and unknown) then f cannot be found from $T_\mu f$.

In these notes we have reported some of the relatively recent results dealing with the Attenuated Radon Transform. I hope that the reader is encouraged to contribute to this field since the state of the art is relatively primitive and there are many unsolved problems some of which are probably easy and may lead to important results.

BIBLIOGRAPHY

1. S. Bellini et al., "Compensation of Tissue Absorption in Emission Tomography", Trans. IEEE ASSP-27:213-218 (1979).

2. T. Budinger, G.T. Gullberg, and R.H. Huesman, "Emission Computed Tomography", in G. Herman, ed., Image Reconstruction from Projections: Implementation and Applications, Topics in Applied Physics Vol. 32, Springer (1979).

3. Y. Censor, D.E. Gustavson, A Lent, H. Tuy, "A New Approach to the Emission Computerized Tomography Problem: Simultaneous Calculation of Attenuated and Activity Coefficients", Trans. IEEE NS-26:2775-2779 (1979).

4. G.T. Gullberg, "The Attenuated Radon Transform: Theory and Application in Medicine and Biology", Lawrence Berkley Lab Report LBL-7486 (1979).

5. C. Hamaker, K.T. Smith, D.C. Solomon, and S.L. Wagner, "The Divergent Beam X-Ray Transform", Rocky Mountain Journal of Mathematics, 10:253-283 (1979).

6. D.E. Kuhl and R.Q. Edwards, "Image Plane Separation Radioisotope Scanning", Radiology, 80:653-662 (1963).

7. F. Natterer, "The Identification Problem in ECT", in Herman and Natterer, ed., Mathematical Aspects of Computerized Tomography, Lecture Notes in Medical Informatics, Vol. 8, Springer, (1981).

8. G.N. Ramachandran and A.V. Lakshminarayanan, Proc. Nat. Acad. Sci., U.S.A., 68:2236-2240 (1971).

9. O. Tretiak and P. Delaney, "The Exponential Convolution Algorithm for Computed Axial Tomography", A Review of Information Processing in Medical Imaging, ORNL/BCTIC2, Oak Ridge National Laboratory, Oak Ridge, Tennessee, (1978).

10. O. Tretiak and C. Metz, "The Exponential Radon Transform", SIAM, J. Appl. Math., 39:341-354 (October 1980).

DEPARTMENT OF ELECTRICAL AND COMPUTER ENGINEERING
DREXEL UNIVERSITY
PHILADELPHIA, PENNSYLVANIA 19104

Proceedings of Symposia in Applied Mathematics
Volume 27, 1982

COMPUTED TOMOGRAPHY: SOME HISTORY AND RECENT DEVELOPMENTS

A.M. Cormack

My early papers [7,8] on what is now called computed tomography were entitled "Representation of a Function by its Line Integrals, with some Radiological Applications", and in one of them I wrote "One would think that this problem would be a standard part of the nineteenth century mathematical repert›ire, but the author has found no reference to it in standard works". I have still found no nineteenth century reference to the problem, but in the years since 1971 I have come across a number of early discoveries of Radon's problem in the physical sciences, which I would like to share with you since they are not generally known. For more mathematical references, see John [21] and Helgason [19]. In addition, I will sketch my connection with the problem and give some results which I have recently obtained.

The first person I know of who tackled Radon's problem was the great Dutch physicist H.A. Lorentz. He found a solution to the three dimensional problem where a function is to be recovered from its integrals over planes. If $\hat{f}(p,\vec{n})$ is the integral of f over a plane perpendicular to the vector $p\vec{n}$ from the origin, and a distance p from the origin, then f at the origin is given by

$$f(0) = -\frac{1}{8\pi^2}\int \left(\frac{\partial^2 \hat{f}(p,\vec{n})}{\partial p^2}\right)_0 d\omega_n$$

where $d\omega_n$ is an element of solid angle in the direction of $\vec{n}$. Since the origin may be arbitrarily chosen, the result holds for any point. We have no idea why Lorentz thought of the problem, or what his method of proof was, and the only reason we know of his work is that the above result was attributed to him by Bockwinkel [4] who used the result in a long paper on the propagation of light in biaxial crystals.

Following Lorentz came Radon's paper, but since all of you have copies of it, I shall not comment on it.

1980 Mathematics Subject Classification 01A60, 44A15, 45A05

0160-7634/82/0000-0876/$03.00

The problem appears again in Holland in 1925 and again involves the physicists Paul Ehrenfest and George Uhlenbeck (who is still alive in this country). In Uhlenbeck's paper he says that Ehrenfest drew his attention to the result of Lorentz and suggested that he generalize it to n-dimensions, which he proceeded to do using Fourier techniques [33]. No reason for solving the problem was given, and my guess is that it was just a cute problem crying out for a solution. I am indebted to Alberto Grünbaum for drawing my attention to the Bockwinkel and Uhlenback papers.

1936 was a vintage year for it contained two independent discoveries of Radon's problem. One was in Stockholm where Cramér and Wold [15] asked the following question: given all marginal distributions of a probability distribution, can one infer the distribution itself? Marginal distributions are what people in computed tomography call projections or views, so the question is just Radon's problem, and in solving it, Cramér and Wold again used the Fourier integral approach. The second was in Leningrad where the well-known Armenian astronomer, V.A. Ambartsumian, posed the following question which he attributes to Eddington [3]. If one looks in a particular direction in space, one sees many stars and these are moving about relative to one another and to the sun. Astronomers would like to know their distribution of velocities but, except for a few close stars, one can only measure their radial velocities, which are deduced from the Doppler shifts of their spectra. The problem then is to deduce the actual distribution of velocities in three dimensions in space from the distributions of radial velocities in various directions. (It is, of course, necessary to assume the velocity distribution to be invariant under spatial translations.) This is just Radon's problem in a three dimensional velocity space rather than ordinary space, and Ambartsumian gave the solution in two and three dimensions [1] in the same form as Radon. Furthermore, he took groups of stars of three different spectral types, with four to five hundred stars in each group, and he used his theoretical results to deduce their actual velocity distributions from the distributions of their radial velocities [2]. One of his results is shown in Fig. 1 and it is in fact the velocity distribution projected onto the galactic plane. This is the first numerical inversion of the Radon transform and it gives the lie to the often made statement that computed tomography would be impossible without computers. Details for the calculation are given in Ambartsumian's paper, and they suggest that even in 1936 computed tomography might have been able to make significant contributions to, say, the diagnosis of tumors in the head.

The problems of finding abnormalities in the soft tissues from X-rays, even with the aid of computed tomography, are far greater than most people realize so it seems to me quite possible that Ambartsumian's numerical methods

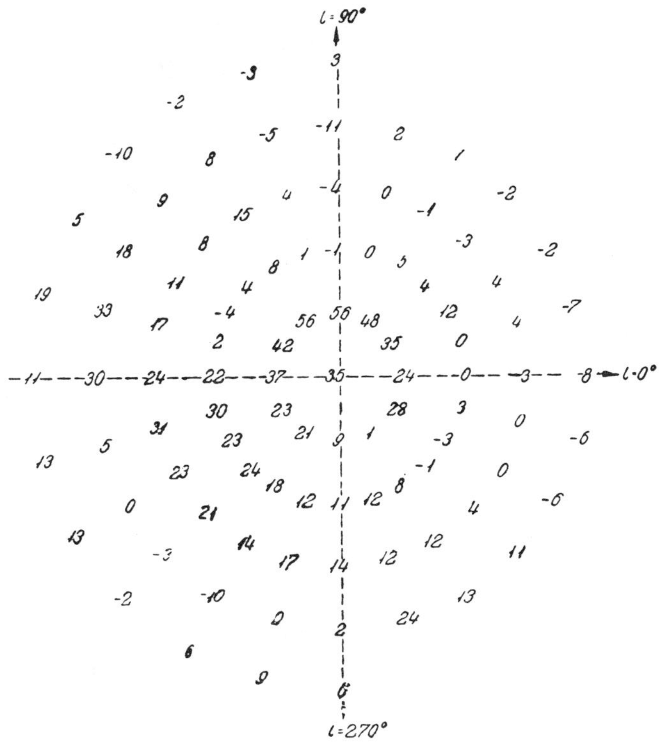

Figure 1

Velocity distribution of stars projected onto the galactic plane.

might have made significant contributions to that part of medicine had they been applied in 1936. I am indebted to Professor Ambartsumian for drawing my attention to his work and for telling me that he was informed of Radon's results two years after he published his work.

In 1947 there is a fascinating abstract by Szarski and Wazewski [32] which begins: "En disposant des fotographies roentgeniques de la tête humaine de tous les côtés, peut-on déterminer la structure intérieure du crâne, voici un problème proposé par M. Stanislaw Majerek, médicin de Cracovie qui, sous un forme élémentaire, a indiqué un méthode de solution qui nécessite une justification mathématique". They then go on to describe Radon's problem and formulate it in terms of a set of "fonctions cylindrique" and state that the problem consists of finding whether this set of functions tends to a solution of Radon's problem. I have been unable to find any further references to this work.

If one looks at the sun, as Bracewell was doing around 1956, with a long radio antenna with a parabolic section perpendicular to its length the signal one gets is an average over a thin strip of the sun of its radio emission at some wavelength. On moving the antenna one gets averages over different strips, and once again we have Radon's problem, which Bracewell proceeded to solve using Fourier and other methods [5].

As I have recounted elsewhere [10] in 1956 I came across the problem in radiology where one has to determine the X-ray absorption coefficient from point to point in a disc from attentuation measurements on X-ray beams traversing the disc; Radon's problem again! Since the treatment I gave differs from previous treatments, and since it shows some interesting features which appear later, I will outline it.

First consider a circularly symmetrical function f(r). Then if $\hat{f}(p)$ is the line integral along a line which is a perpendicular distance p from the origin, it is easy to show that

$$\hat{f}(p) = 2\int_p^\infty \frac{f(r)rdr}{\sqrt{r^2-p^2}} . \qquad (1)$$

This is just Abel's equation with the well-known solution

$$f(r) = -\frac{1}{\pi}\int_r^\infty \frac{\hat{f}'(p)dp}{\sqrt{p^2-r^2}} \qquad (2)$$

which is quite amenable to numerical analysis and which I used in an experiment [7]. The fact that the integral in Eq. (2) extends from r to ∞ is an explicit expression of the hole theorem: to find f(r) one needs only $\hat{f}(p)$ for $p \geq r$. This remains true for non-circularly symmetrical functions as can be seen from the following. Let the line L along which $\hat{f}$ is calculated be specified by (p,ϕ) as in Fig. 2.

Reconstruction of Densities from their Projections

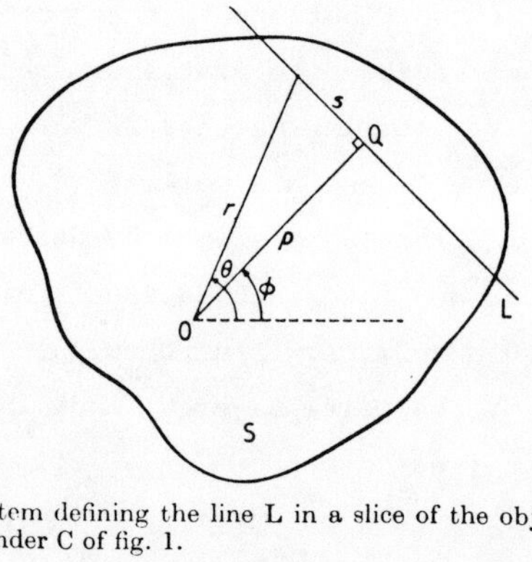

Fig. 2. Coordinate system defining the line L in a slice of the object. The line L is the limit of the cylinder C of fig. 1.

Figure 2.

Coordinates for the Radon transform in R^2

Then if one writes

$$f(r,\theta) = \sum_\ell f_\ell(r)e^{i\ell\theta}, \qquad \hat{f}(p,\phi) = \sum_\ell \hat{f}_\ell(p)e^{i\ell\phi},$$

one can show that

$$\hat{f}_\ell(p) = 2\int_p^\infty \frac{f_\ell(r)T_\ell(p/r)rdr}{\sqrt{r^2-p^2}}, \tag{3}$$

which has the solution

$$f_\ell(r) = -\frac{1}{\pi}\int_r^\infty \frac{\hat{f}_\ell'(p)T_\ell(p/r)dp}{\sqrt{r^2-p^2}}, \tag{4}$$

where $T_\ell(x)$ is the Tschebicheff polynomial of the first kind which is orthogonal in $(-1,1)$ [8]. In Eq. (4), $T_\ell(x)$ appears for $x \geqq 1$, and its leading term is const $\times\ x^\ell$ hence if $\hat{f}_\ell'(p)$ is obtained from noisy data, the noise is propagated badly from the exterior in as one goes to smaller values of r. In a recent application of this method in fusion physics [30] it is shown that Eq. (4) gives bad results for $\ell \gtrsim 10$. Hence Eq. (4) is not good for numerical solutions in medicine where one would require ℓ to be of the order of hundreds. I therefore resorted to radial expansions of f_ℓ and $\hat{f}_\ell$ which take the following form. If

$$f_\ell(r) = r^\ell(1-r^2)^{\lambda-1/2}\ G_k(\ell+\lambda,\ \ell+1,\ r^2), \tag{5a}$$

then

$$\hat{f}_\ell(p) = \text{const} \times\ (1-p^2)^{\lambda-1/2}\ C^\lambda_{\ell+2k}(p), \tag{5b}$$

where the G_k are Jacobi polynomials in $(0,1)$ and the C^λ_ℓ are Gegenbauer polynomials. This expansion, for general λ, was recently given by Louis [22] and I used only the case $\lambda = 1$ which has been discussed by Marr [24], Ein-Gal [17] and others. Application of these formulae in an experiment to find f from $\hat{f}$ showed that computed tomography would work well for real data, but today the most commonly used numerical methods are based on the convolution algorithm of Ramachandran and Laksminarayanan [29].

When Radon's problem is extended to R^n Ludwig [23] and Deans [16] obtained formulae which contain Eqs. (3) and (4) as special cases. For more general spaces, a more general form of the hole theorem is immediately obtainable from a theorem (2.1) of Helgason [18].

Since the middle sixties there has been an explosion in applications of Radon's problem to many problems in science and engineering some of which are given in Marr [25] and more in a hard-to-obtain bibliography by Richard Gordon. Since the advent of commercial computed tomography scanners [20] there has been a greater explosion in the medical field, and reviews of Brooks and DiChiro [6] and Shepp and Kruskal [31] are useful, though both are becoming dated.

Radon's problem in R^2 gives rise to a nice tiling problem which has not been used and which may not even be useful. Consider two pairs of points, S_1,S_2 and S_1',S_2' which define a beam of X-rays traversing a sample in the following way: only lines which intersect S_1S_2 and $S_1'S_2'$ contributed to a measurement of some average value of f with this beam. The lines defined by S_1S_2 and $S_1'S'_2$ are represented by a region of the Radon transform bounded by four circles, namely, those with radii OS_1, OS_2, OS_1', OS_2' where O is the origin of the transform. The question then is how the Radon transform of a finite object can be covered exactly by a set of such beams which intersect it. The solution is easy for a model of positron-annihilation (PET) scanning when the detectors can be considered to be arcs of the circle defining the object [11]. For exactly parallel beams (i.e. S_1S_2 and $S_1'S_2'$ forming the corners of a fixed rectangle) there is no exact solution, but a mix of parallel and almost-parallel beams does provide a solution [9]. One wonders whether there are other solutions, and whether these results can be extended to R^n.

When I had obtained a solution of Radon's problem in the form given in Eq. (4), I asked myself whether there were other curves specified by parameters (p,ϕ) for which Radon's problem could be solved. I immediately tried circles through the origin with diameter p and orientation of the diameter relative to the x-axis given by ϕ, and obtained results remarkably similar to Equations (3) and (4), namely

$$\hat{f}_\ell(p) = 2\int_0^p \frac{f_\ell(r)T_\ell(r/p)dr}{\sqrt{p^2-r^2}}\,, \tag{6}$$

which has the solution

$$f_\ell(r) = \frac{1}{\pi}\,\frac{d}{dr}\int_0^r \frac{f_\ell(p)T_\ell(r/p)dp}{\sqrt{r^2-p^2}}\,. \tag{7}$$

Quinto and I recently generalized these to R^n and applied the results to solutions of Darboux' equation [14]. The results will not be given here, but the kernels are more complicated than those in Eqs. (6) and (7), involving Gegenbauer polynomials which appear here as in [23] and [16] as a result of using the Funk-Hecke theorem for spherical harmonic expansions.

As an older physicist, I was aware that in electrostatics straight lines and planes are related to circles and spheres by inversion $((r,\theta) \to (1/r,\theta))$ and this connection intrigued me for a long time. Recently I came across two families of curves which are related by inversion, and for which Radon's problem can be solved, which include the straight lines and circles. These are, in the previous notation,

$$\begin{aligned} r^\alpha \cos\{\alpha(\theta-\phi)\} &= p^\alpha,\ \alpha > 0 \\ p^\beta \cos\{\beta(\theta-\phi)\} &= r^\beta,\ \beta > 0. \end{aligned} \tag{8a}$$

For $\alpha = 1/2,\ 1,\ 2$ we have parabolae, straight lines and hyperbolae respectively, and for $\beta = 1/2,\ 1,\ 2$ we have cardioids, circles through the origin and one branched lemniscates of Bernoulli, respectively. Many of these curves are the orbits of particles in central fields or of charged particles in certain magnetic fields. The details can be found in [12], but in general the results are very like those for the ordinary Radon transform: the inversion formulae are similar and the appropriate hole theorems make their appearance. More recently [13] certain other properties have been obtained. For general values of α and β certain inverse and direct powers of r ($r \neq 0$) yield null Radon transforms, as pointed out by Newman [26] and Louis [22] for the ordinary Radon transform. For $\alpha = 1$, $\beta = 1$, Quinto [28] has shown that these constitute the entire null space. For α or $\beta = 1/m$, m = 1, 2, 3... a number of orthogonal expansions of f_ℓ and $\hat{f}_\ell$ are given. For α-curves in (0,1) the expansions are an extension of the results of Louis (Eqs. 5a,5b); in $(1,\infty)$ they are an extension of Perry's results [27]. In $(0,\infty)$ they follow the pattern given in [8]. For β-curves, the inversion property allows one to write down immediately the appropriate formulae for (0,1) and $(1,\infty)$ from the results for α-curves for $(1,\infty)$ and (0,1) respectively.

Working on Radon's problem has given me a great deal of pleasure and, as usual, each question answered has raised new questions. There are still plenty of these to be answered.

BIBLIOGRAPHY

1. Ambartsumian, V., M.N.R.A.S., 96 (1936), 172.
2. Ambartsumian, V., Publ. de L'Obs. Astron. Leningrad 7 (1936) 22.
3. Ambartsumian, V., Ann. Rev. Astron. Astro. 18 (1980) 1.
4. Bockwinkel, H.B.A., Verh. Kon. Akad. Wet. Wissen. Naturk. 14 (1906) 636.
5. Bracewell, R.N., Austr. J. Phys. 9 (1956) 198.
6. Brooks, A.R., and DiChiro, G., Phys. Med. Biol. 21 (1976) 689.
7. Cormack, A.M., J. Appl. Phys. 34 (1963) 2722.
8. Cormack, A.M., J. Appl. Phys. 35 (1964) 2908.
9. Cormack, A.M., Phys. Med. Biol. 23 (1978) 1141.
10. Cormack, A.M., Science 209 (1980) 1482.
11. Cormack, A.M., Phys. Med. Biol. 25 (1980) 543.
12. Cormack, A.M., Proc. Am. Math. Soc. 83 (1981) 325.
13. Cormack, A.M., to appear in Proc. Am. Math. Soc.
14. Cormack, A.M., and Quinto, E.T., Trans. Am. Math. Soc. 260 (1980) 575.
15. Cramér, H. and Wold, H., J. Lond. Math. Soc. 11 (1936) 290.
16. Deans, S., J. Math. Phys. 19 (1978) 2346.
17. Ein-Gal, M., Stanford U. Information Systems Laboratory, Technical Report 6851-1 (1974).
18. Helgason, S., Acta Math. 113 (1965) 153.
19. Helgason, S., The Radon Transform, Birkhauser, Boston (1980).
20. Hounsfield, G.N., Br. J. Radiol. 46 (1973) 1016.

21. John, F., Plane Waves and Spherical Means, Interscience, New York (1955).

22. Louis, A.K., SUNY, Buffalo, Medical Image Processing Report T52 (1981).

23. Ludwig, D., Comm. Pure Appl. Math. 69 (1966) 49.

24. Marr, R.B., J. Math. Anal. Appl. 45 (1974) 357.

25. Marr, R.B., Brookhaven National Laboratory, BNL 20425 (1974).

26. Newman, D.J., cited in reference 31.

27. Perry, R.M., J. Math. Anal. Appl. 59 (1977) 324.

28. Quinto, E.T., to appear in J. Math. Anal. Appl..

29. Ramachandran, G.N. and Laksminarayanan, A.V., Proc. Math. Acad. Sci. U.S.A., 68 (1971) 2236.

30. Sauthoff, N.R. and von Goeler, S., I.E.E.E. Trans. on Plasma Science 7 (1979) 141.

31. Shepp, L.A. and Kruskal, J.B., Am. Math. Monthly 85 (1978) 420.

32. Szarski, J., and Wazewski, T., Ann. Soc. Polon. de Math. 20 (1947) 389.

33. Uhlenbeck, G.E., Physica 5 (1925) 423.

DEPARTMENT OF PHYSICS
TUFTS UNIVERSITY
MEDFORD, MASSACHUSETTS 02155

Proceedings of Symposia in Applied Mathematics
Volume 27, 1982

THE LIMITED ANGLE RECONSTRUCTION PROBLEM

F. Alberto Grünbaum[1]

PREFACE. These lecture notes are intended as a smooth introduction to the subject presented from a particular point of view. They are not written for the experts. A number of open, and sometimes only vaguely stated, problems are included in the text. It is a pleasure to record here my debt to my colleagues as well as to my former students. A good part of these notes is an account of joint work with one of them, M. Davison. The reader may find useful, besides the other lectures in this short course, the review by R. Marr [1] and the book [2] edited by G. Herman.

TABLE OF CONTENTS

1. INTRODUCTION

The purpose of these notes is to discuss the problem of recovering $f(\underline{x})$, $\underline{x} \in R^n$, from a finite collection of its one dimensional projections

$$Pf(\omega_j, t) \equiv \int_{\langle \underline{x}, \omega_j \rangle = t} f \equiv \int_{R^n} f(\underline{x})\ \delta(t - \langle \underline{x}, \omega_j \rangle)$$

$$\omega_j \in S^{n-1} \qquad j = 1, \ldots, M$$

Here S^{n-1} denotes the unit sphere in R^n, we use $\langle \underline{x}, \omega_j \rangle$ for the usual scalar product in R^n, and δ denotes Dirac's delta function.

---

[1] Supported by NSF grant MCS-81-07086

0160-7634/82/0000-0877/$05.75

The most common applications call for $n = 2$ or 3, and from time to time we specialize to these two cases. Many applications however cannot be put in the form given above: a good example is given by fan or cone beam tomography.

We will be concerned with the effect that different choices of directions ω_j have on the reconstruction of f from its projections $Pf(\omega_j, t)$.

We make the very convenient -- and plausible -- assumption that we are dealing with "complete projections", i.e. that $Pf(\omega_j, t)$ is known for all values of t. This allows us to focus attention on the choice of ω_j. We denote such a projection by $Pf(\omega_j)$.

However in analyzing the optimal use of a <u>finite</u> number of measurements

$$Pf(\omega_j, t_k) \quad j = 1, \ldots, M; \; k \in K_j \text{(finite)}$$

one should also study the trade-off between the "angular" and the "spatial" sampling of the projections. Problems like this become very hard to handle: we offer one such

PROBLEM 1. Given the number N find the "optimal" choice of N hyperplanes

$$\langle \underline{x}, \omega_j \rangle = t_k$$

for the problem of recovering an unknown function f from its N integrals

$$Pf(\omega_j, t_k) \; .$$

The number of hyperplanes need not be the same for each direction: the only requirement is

$$\sum_{j=1}^{M} \text{number of elements in } K_j = N$$

This problem is not quite posed unless we specify something about the class of functions f and our meaning of "optimal". But doing this, in a reasonable way, is part of the problem.

From now on we assume that we are dealing with "complete projections" as defined above. The difficulty that remains has to do with <u>discretization</u> in ω. Indeed if one knew $Pf(\omega, t)$ for all values $\omega \in S^{n-1}$ and all $t \in R$, then one could use the formulas of Radon as given in many standard references, see [3], [4], and [5].

In the 2-dimensional case we have

$$f(x,y) = \frac{1}{2\pi^2} \int_0^{\pi} d\theta \; \text{p.v.} \int_{-\infty}^{\infty} \frac{1}{t - s} \frac{\partial}{\partial s} Pf(\theta, s) \, ds$$

with $t = x \cos \theta + y \sin \theta$.

This expression illustrates the main features of the formula for any dimension: The projections are first "filtered" by the application of some integral (or differential) operator. Then, for each fixed ω in S^{n-1}, the resulting function of the variable t is "back projected" to get a function of $\underline{x} = (x,y)$: this means simply that the function is made constant along the hyperplanes $\langle \underline{x},\omega \rangle = t$.

Finally all these back projections are added up with uniform weight over S^{n-1}. The first step can be rewritten in the form of a "convolution" of $Pf(\theta,\cdot)$ with a filter function $\alpha(\cdot)$

$$-\int_{-\infty}^{\infty} \frac{\partial}{\partial s} \frac{1}{t-s} Pf(\theta,s)ds \equiv (\alpha * Pf(\theta))(t)$$

The nature of α depends on the dimension n in R^n: α is a global operator if n is even and a local one if n is odd. In the case of $n = 3$ we get

$$(\alpha * Pf(\theta))(t) = \frac{d^2}{dt^2} Pf(\theta,t)$$

To make the inversion formula valid one has to assume some technical conditions on $f(\underline{x})$, see [3,4,5].

Radon's formula is a nice example of an important tool that is rediscovered over and over again by workers in different areas. While it seems that Radon was the first one to prove the n-dimensional case in 1917, the Dutch physicist H. Lorentz knew the formula in the 3 dimensional case and made it part of his lectures around the end of last century, [6]. A paper in 1906 credits the result to Lorentz [7]. Several years later Ehrenfest was interested in the different behaviour of the wave equation in odd and even dimensional spaces and proposed to Uhlenbeck that he should take up extending Lorentz formula. And thus in the same year of his celebrated paper on the spin of the electron (1925), Uhlenbeck unaware of the work of Radon obtained the inversion formulas for arbitrary dimension, [8].

Several years later Cramer and Wold [9], Bracewell [10] and then Cormack [11] had to rediscover this tool on their own, and in the case of Cormack put it to use in our present context. Another independent attack on the problem is due to Hounsfield [12]. Our main interest in this formula is that it suggest what to do when one is using only a finite number of complete projections.

Indeed if we have M directions ω_j and their corresponding one dimensional projections $Pf(\omega_j,t)$, we propose to get an approximation to $f(\underline{x})$ in the form

$$\tilde{f}(\underline{x}) = \frac{1}{M}\sum_{i=1}^{M} (Pf(\omega_j) * \alpha_j)(\langle \underline{x},\omega_j\rangle)$$

If ω_j are given, the main task is to determine the filter functions α_j so that the "difference" between $\tilde{f}(\underline{x})$ and $f(\underline{x})$ be "small".

Before we start, it is proper to gain some perspective: results of K. Smith et. al. [13] indicate that a finite set of projections cannot really serve to approximate $f(\underline{x})$ and yet, a finite number of line or plane integrals is all that one can feed into any reconstruction algorithm. Assuming that the reconstructions obtained in clinical and other applications bear some connection with the true object we have a paradox.

A way out of this seeming paradox is not hard to find. It is the basis of the development in [14]. One gives up trying to reconstruct pointwise values of $f(\underline{x})$ -- on the basis that if one had all this information one would not know what to do with it -- and settles for the task of obtaining filters α_i such that $\tilde{f}(\underline{x})$ is close to a <u>blurred or mollified version</u> of f, i.e. $f*\phi$, with ϕ is an "approximate delta function of spread σ". As long as one can do this for any choice of "spread σ" -- notice that the filters α_i will depend on σ -- one should try to prove that for a large enough number of directions $M(\sigma)$ the discrepancy

$$\tilde{f}(\underline{x}) - (f*\phi_\sigma)(\underline{x})$$

can be made small once some <u>a priori information</u> about f is provided.

This way of "regularizing" the problem has been explicitly or implicitly used before, among others by Cho [15] in tomography and by Backus and Gilbert [16] in geophysics.

2. ERROR BOUND FOR FILTERED BACK PROJECTION METHODS

The issue here is how to measure the discrepancy between

$$\tilde{f}(\underline{x}) = \frac{1}{m}\sum_{i=1}^{M} (Pf(\omega_j)*\alpha_j)(\langle \underline{x},\omega_j\rangle)$$

and the "blurred version" of f, namely $f*\phi$, once ω_j, α_j and ϕ have (somehow) been chosen.

No natural concept comes forward except the vaguely defined notion that we should produce an $\tilde{f}(\underline{x})$ which has a great deal of "diagnostic value": for instance one may be willing to accept a distorted picture of f as long as some kind of fine details in the original f are not erased; such an approximation to f will not be "close" to f in any standard sense.

At this point we take the mathematical way out, and settle for the supremum norm of the difference between $\tilde{f}$ and $f*\phi$ as an indication of the error. This is really a question of expediency since other mathematically acceptable norms could be used at this point. This point about expedience may be formulated as a

PRINCIPLE: Norms, and other mathematical constructs, should be judged by the effectiveness and quality of the algorithms they lead to.

Having settled on the sup norm, one can easily derive the "error bound"

$$\sup_{\bar{y}\in\frac{1}{2}D} |\tilde{f}(\bar{y}) - (f*\phi)(\bar{y})| \le \sup_{\bar{y}\in\frac{1}{2}D} \|f_{\bar{y}}\|_w$$

$$\times \|\phi - \frac{1}{M}\sum_{j=1}^{M} B_j\alpha_j\|_{1/w}$$

Here a number of terms need defining: D is the unit ball centered at the origin in R^n, f a real valued function with support in $\frac{1}{2}D$: the ball centered at the origin and of radius $\frac{1}{2}$, $f_{\bar{y}}$ the translate of f by the vector $\bar{y}$, $w(\underline{x})$ an arbitrary positive measurable weight function supported in D, $\|h\|_w$ means the $w(\underline{x})$ weighted L^2 norm

$$\|h\|_w = \left\{\int_D h^2(\underline{x})\, d\underline{x}\right\}^{1/2} .$$

Finally B_j is defined for each ω_j by

$$(B_j\alpha_j)(\underline{x}) \equiv \alpha_j(\langle \underline{x},\omega_j\rangle)$$

The proof of this bound uses just the Cauchy-Schwartz inequality. Moreover it is clear that even if we stick to the supremum norm on the left hand side of our inequality, we are not confined to picking the w and $1/w$ weighted norms on the right hand side. Any pair of dual norms will do, for instance the Sobolev spaces H_α, $H_{-\alpha}$ could replace L^2_w and $L^2_{1/w}$.

The point is that if we have a priori information about f in the form of

$$\sup_{\bar{y}\in\frac{1}{2}D} \|f_{\bar{y}}\|_w < \mathbb{C}$$

then assuming ω_j, w and ϕ given, the choice of good filters α_j is reduced to picking them so that the second factor in the error bound above will be as small as possible.

One of the main troubles with the approach discussed here is that very limited use is made of a-priori information about f. This gives us the

PROBLEM 2. Modify the approach given in these notes to better incorporate extra knowledge about the unknown $f(\underline{x})$.

3. CHOICE OF FILTERS AND WEIGHT

Consider the problem of choosing, once ω_j, w and ϕ are given, filter functions α_j so as to minimize the expression

$$\|\phi - \frac{1}{M} \sum_{j=1}^{M} B_j\alpha_j\|_{1/w}$$

This problem can be put in familiar geometrical language. Indeed take w_1 to be an arbitrary positive weight in the interval $I = [-1,1]$ and consider the map

$$P : L^2(D,w) \to \bigoplus_{j=1}^{M} L^2(I,w_1)$$

given by

$$Pf = (Pf(\omega_1), \ldots, Pf(\omega_M))$$

The inner product in $\bigoplus_{j=1}^{M} L^2(I,w_1)$ for two M-tuples of functions $\beta = (\beta_1, \ldots, \beta_M)$, $\beta_j \in L^2(I,w_1)$ is given by

$$\langle \beta,\gamma \rangle = \frac{1}{M} \sum_{j=1}^{M} \int_{-1}^{1} \beta_j(s)\, \gamma_j(s)\, w_1(s)\, ds$$

The adjoint, P^t, of P satisfies

$$P^t\beta = \frac{1}{M} \sum_{j=1}^{M} P^t\beta_j(\omega_j)$$

where we set, for each j,

$$P^t g(\omega_j) = g(\langle \underline{x},\omega_j \rangle)\, \frac{w_1(\langle \underline{x},\omega_j \rangle)}{w(\underline{x})}$$

It is now straight forward that minimizing the expression at the beginning of this section is equivalent to minimizing

$$\|\nu - P^t\beta\|_w$$

as long as we set

$$\nu(\underline{x}) = \phi(\underline{x})/w(\underline{x})$$

and

$$\beta_j(x) = \alpha_j(x)/w_1(x) \qquad j = 1, \ldots, M$$

This minimization problem is solved by solving the corresponding "normal equations"

$$PP^t\beta = P\nu$$

This means that we choose β so that although $P^t\beta$ is not quite ν, the projections of these two functions on D along the directions ω_j $j = 1, \ldots, M$ should agree.

Writing out the normal equations in full we get

$$P\left(\frac{1}{M}\sum_{i=1}^{M} P^t\beta_i(\omega_i)\right)(\omega_j) = P\nu(\omega_j) \qquad j = 1, \ldots, M.$$

This system of equations could be attacked numerically by replacing $P\nu(\omega_j)$ by its expansion in terms of some orthogonal basis of functions in $L^2(I,w_1)$ and keeping the first N terms in this expansion. This would allow the determination of the first N terms of the expansion of each β_i in this basis: a system of $M \times N$ equations in $M \times N$ unknowns.

Any realistic situation will require $M \cong 200$ and $N \cong 500$ -- even higher values of N are needed if σ the "spread of ϕ" is small enough -- and thus the solution of the normal equations cannot be pushed any further since we would be dealing with a <u>full</u> and extremely <u>large</u> matrix. But a <u>miracle</u> comes to our rescue: if $w(x)$, $w_1(x)$ are chosen properly, a basis can be found in $\bigoplus_{i=1}^{M} L^2(I,w_1)$ such that we get a block diagonal matrix problem with blocks of size $M \times M$. We are still unable to give an analytic solution -- except in special cases -- but now we have a sequence of manageable numerical problems which can -- say for $M \leq 200$ -- be easily handled with existing computers.

The normal equations, can be written, at least if $w(\underline{x})$ is <u>rotationally invariant</u> in the form

$$\frac{1}{M}\sum_{i=1}^{M} A(\langle\omega_i,\omega_j\rangle)\beta_i = P\nu(\omega_j)$$

Here $A(\langle\omega_i,\omega_j\rangle)g(t) \equiv P(P^t g(\omega_i))(\omega_j)(t) = \displaystyle\int_{\langle x,\omega_j\rangle = t} \frac{g(\langle\underline{x},\omega_i\rangle)w_1(\langle\underline{x},\omega_i\rangle)}{w(\underline{x})}\,d\underline{x}$

and this function of t is seen to depend only on $\langle\omega_i,\omega_j\rangle$ by picking ω_i', ω_j' so that

$$\langle \omega_i', \omega_j' \rangle = \langle \omega_i, \omega_j \rangle$$

and bringing in a rotation O in R^n such that $O^*\omega_i = \omega_i'$, $O^*\omega_j = \omega_j$. Then we get

$$P(P^t g(\omega_i'))(\omega_j')(t) = \int_{\langle O\underline{x}, w_j \rangle = t} \frac{g(\langle O\underline{x}, \omega_i \rangle) w_1(\langle O\underline{x}, \omega_i \rangle)}{w(\underline{x})} d\underline{x}$$

and a change of variables from $\underline{x}$ to $O\underline{x}$ does the job.

We can now state the main result of this section (formulated here only for $n = 2$ for simplicity).

THEOREM. If $w(\underline{x}) = \frac{\pi}{\lambda}(1 - |x|^2)^{1-\lambda}$ $\lambda > 0$ and

$$w_1(x) = (1 - x^2)^{\frac{1}{2} - \lambda} \frac{\pi^{\frac{1}{2}}\Gamma(\lambda + 1/2)}{\lambda\Gamma(\lambda)}$$

then the family of operators

$$A(\langle \omega_i, \omega_j \rangle) : L^2(I, w_1) \to L^2(I, w_1)$$

defined by

$$A(\langle \omega_i, \omega_j \rangle) f = P(P^t f(\omega_i))(\omega_j)$$

is commutative with a common set of eigenfunctions given by

$$\phi_m(t) = \mathbb{C}_m(t)/w_1(t)$$

and eigenvalues given by

$$\mathbb{C}_m^\lambda(\langle \omega_i, \omega_j \rangle)/\mathbb{C}_m^\lambda(1)$$

Here $\mathbb{C}_m^\lambda$ denotes the m-th Gegenbauer or ultraspherical polynomial, [17].

NOTE: These results are proved in detail in [18] and we omit the proof here. For the case of $\lambda = 1$ these results are already contained in Logan and Shepp [19] and Hammaker and Solmon [20]. There is no reason for taking D as the unit disk: if the radius is taken to be R everything works with the obvious changes: one encounters $\mathbb{C}_n^\lambda(x/R)$ instead of $\mathbb{C}_n^\lambda(x)$. Finally if in the manner of Maxwell, Poincare and others one sets $R = \sqrt{\lambda}$ and exploits the true meaning of the relation

$$\lim_{\lambda \to \infty} \lambda^{n/2} \mathbb{C}_n^\lambda\left(\frac{x}{\lambda^{\frac{1}{2}}}\right) = \frac{H_n(x)}{n!}$$

with $H_n(x)$ the Hermite polynomial, one can see that the theorem above holds for $L^2(R^n, e^{|x|^2})$. Details can be seen in [18]. See also [21,22].

The importance of this theorem is that it allows one to analyze the action of PP^t on $\bigoplus_{i=1}^{M} L^2(I,w_1)$ in terms of its action on a sequence of orthogonal invariant subspaces. For details see Corollary 3.1 and Lemma 3.1 in [18]. The solution is presented in section 4 of these notes.

We see now that given the directions ω_i and the "desired point response function ϕ", we cannot pick $w(\underline{x})$ arbitrarily if we want to effectively solve the normal equations giving the optimal choice of filters α_i.

One may wonder if the weights $w(\underline{x})$ given above -- a one parameter family depending on λ-exhaust the possibilities leading to a computationally simple situation. The answer is almost surely yes: the commutativity property of the theorem above characterizes the Gegenbauer and Hermite weights. A proof is given in [18].

Notice that since the only condition on λ is $\lambda > 0$ we can play with weights which penalize more the boundary of D than its center or just the contrary. For each weight in the family a reconstruction is possible and we can now recall the principle announced earlier: judge the right choice of λ by looking at reconstructions done with different values of λ.

We can now formulate the

PROBLEM 3. Find other weights, besides those given above, so that for arbitrary directions ω_j the normal equations can be handled numerically by a proper choice of basis in $\bigoplus_{i=1}^{M} L^2(I,w_1)$.

4. THE NORMAL EQUATIONS AND AN EXPRESSION FOR THE ERROR BOUND

The effective solution of the normal equations is achieved using the functions $\phi_m(t)$ introduced in section 3.

Put

$$\beta_i(t) = \sum_{m=0}^{\infty} a_m^i \phi_m(t)$$

and

$$(P\nu)(\omega_j)(t) = \sum_{m=0}^{\infty} b_m^j \phi_m(t)$$

Using the facts mentioned in section 3 we see that we have to solve an infinite set of $M \times M$ matrix problems which determine the coefficients a_m^i, namely

$$\frac{1}{M} \sum_{i=1}^{M} \frac{\mathbb{C}_m^\lambda(\langle \omega_i,\omega_j \rangle)}{\mathbb{C}_m^\lambda(1)} a_m^i = b_m^j, \; 1 \leq j \leq M.$$

To conclude that these formal manipulations really give a solution to

$$PP^t\beta = P\nu$$

for any ν in $L^2(D,w)$ we have to show that the range of P^t is a closed subspace of $L^2(D,w)$. This has been done in several places in the literature: it boils down to showing that the nonzero eigenvalues of PP^t are bounded away from zero, see [18] and for $\lambda = 1$ see [20].

Picking $\beta = (\beta_1, \beta_2, \ldots, \beta_M)$ in the fashion discussed above has the effect of rendering the quantity

$$\|\nu - P^t\beta\|_w$$

and therefore (see section 3) the expression

$$\|\phi - \frac{1}{M}\sum_{i=1}^{M} \beta_j\alpha_j\|_{1/w}$$

as small as it is possible, once $w(\underline{x})$, $\phi(\underline{x})$ and the directions ω_i are fixed.

Since the corresponding functions $\alpha_1, \ldots, \alpha_M$ given by

$$\alpha_j(t) = \beta_j(t)w_1(t)$$

are to be used as filters in the reconstruction, any errors in the values of $P(\omega_j)f$ get badly magnified if $\alpha_j(t)$ is "large" in some sense. This leads, through an analysis which can be made quite rigorous, see [18], to the conclusion that it may be advantageous to minimize the expression

$$\|\nu - P^t\beta\|^2 + \varepsilon\|\beta\|^2$$

Here ε is a measure of the variance of the noise in the measurements $P(\omega_j)f$. Setting $\varepsilon > 0$ brings in a compromise between good reconstructions with clean data and well conditioned reconstructions with noisy data.

The analysis to follow can be easily changed when $\varepsilon > 0$. We take $\varepsilon = 0$. We give below an explicit formula which expresses the minimum possible value for the error bound in section 2, namely

$$\|\phi - \frac{1}{M}\sum B_j\alpha_j\|^2$$

This will be done when ϕ is taken to be a <u>radial</u> function, and for convenience $\lambda = n/2$. Moreover we only look at $n = 2$ and 3.

First some notation

C_m = dimension of the space of homogeneous polynomials of degree m in n variables

$= m + 1$ for $n = 2$ and $\frac{(m+2)(m+1)}{2}$ for $n = 3$

$A_m = M \times M$ matrix whose i,j entry is given by $\frac{\mathbb{C}_m^\lambda(\langle \omega_i, \omega_j \rangle)}{\mathbb{C}_m^\lambda(1)}$

Finally put

$$b_m = \begin{cases} e^{-(m+2)m\sigma} & \text{for } n = 2 \\ \frac{3}{4\pi} e^{-(m+3)m\sigma} \frac{(m+1)!!}{m!!} \sqrt{\frac{m!(3/2+m)!}{(3)_m}} & \text{for } n = 3 \end{cases}$$

and denote the vector $(1, 1, \ldots, 1)^T$ by $\mathbb{1}$. Then the promised formula is

$$\sum_{m=0}^{\infty} b_{2m}^2 \left(\frac{1}{C_{2m}} - (A_{2m}^{-1} \mathbb{1}, \mathbb{1}) \right)$$

For this formula for the best error bound for the error to hold we are taking $\phi(\underline{x}) \equiv \phi(\underline{x},\sigma)$ to be the solution at "time" σ of an appropriate "heat equation" with $\phi(\underline{x},0) = \delta(\underline{x})$. For details see [18]. With this choice we get

$(P\nu)(\omega_j)$ = independent of ω_j because of the radial assumption

$$= \sum_{m=0}^{\infty} b_{2m}\phi_{2m}$$

The matrices A_{2m} are nonnegative symmetric. Moreover one can see that

$$\frac{1}{C_{2m}} \geq (A_{2m}^{-1} \mathbb{1}, \mathbb{1})$$

and thus the smallest possible upper bound for the (square of the) error given above is a sum of nonnegative terms representing the square of the error in approximating $\phi(\underline{x})$ by a sum of back projections, as a sum of the squares of the errors in each one of the orthogonal subspaces of L^2 made up of homogeneous polynomials of different degrees.

The formula given above will be used in section 5 as a guide in determining "good choices" of directions.

We close this section with the

PROBLEM 4. Extend the expression given above in the case when $\phi(\underline{x})$ is not assumed to be radial.

5. CHOICE OF DIRECTIONS AND OF MOLLIFIER

The best possible bound for the error given in section 4 suggests a way to pick directions $\omega_1, \ldots, \omega_M$: pick them so that

$$\sum_{m=0}^{\infty} b_{2m}^2 (A_{2m}^{-1} \mathbf{1}, \mathbb{1})$$

be as large as possible.

We remarked earlier that each term $(A_{2m}^{-1} \mathbf{1}, \mathbf{1})$ is bounded independently of the choice of directions. Now we will show that this upper bound is achieved as soon as the number of directions M is large enough compared to m. First we need the

THEOREM. Let $\omega_1, \ldots, \omega_M$ be unit directions in the sphere such that an exact quadrature formula of the type

$$\int_{|\omega|=1} F(\omega)\, d\omega = \sum_{i=1}^{M} \alpha_j F(\omega_j)$$

holds for all (even) polynomials in (x,y,z) of degree $\leq k$. If $f(\xi)$, $\xi \in R$, is any (even) polynomial of degree $\leq k$, and the $M \times M$ matrix A is given by

$$A_{ij} = f(\langle\langle \omega_i, \omega_j \rangle\rangle)$$

we have

$$(A^{-1} \mathbb{1}, \mathbb{1}) = \frac{1}{\int_{|\omega|=1} f(\langle\langle v, \omega \rangle\rangle)\, d\omega}$$

with v any unit vector.

PROOF. For each ω_i, $i = 1, \ldots, M$ the function

$$F_i(u) = f(\langle\langle \omega_i, u \rangle\rangle)$$

is an (even) polynomial of degree $\leq k$. Each of these functions has the same integral on S_2, since the measure on S_2 is rotation invariant. Thus

$$\sum \alpha_i F_i(\omega_j) = \int_{|u|=1} f(\langle\langle \omega_i, u \rangle\rangle)\, du = \text{independent of } i.$$

But this means that if

$$\alpha = (\alpha_1, \alpha_2, \ldots, \alpha_M)$$

then

$$A^{-1} \mathbb{1} = \frac{\alpha}{\int_{|u|=1} f(\omega_i, u)\, du}$$

and since, by normalizing the measure, we have $\sum \alpha_i = 1$, one concludes that

$$(A^{-1}\mathbb{1}, \mathbb{1}) = \frac{1}{\int_{|\omega|=1} f(\langle v,\omega\rangle)\, d\omega}$$

as advertised above.

Now we apply this to the unit sphere in dimension three to obtain

COROLLARY. Given M directions with

$$M \geq (k+1)(2k+1)$$

then

$$\frac{1}{C_{2k}} = (A_{2k}^{-1}\mathbb{1}, \mathbb{1})$$

PROOF. Direct evaluation shows that

$$\int_{|u|=1} \frac{\mathbb{C}_{2n}^{3/2}(\langle v,u\rangle)\, du}{\mathbb{C}_{2n}^{3/2}(1)} = \frac{1}{(2n+1)(n+1)}$$

Observe that, since we are taking for simplicity $\varepsilon = 0$, $A^{-1}\mathbb{1}$ is not uniquely defined. However the expression $(A^{-1}\mathbb{1}, \mathbb{1})$ is well defined.

The result given above indicates that if $\omega_1, \ldots, \omega_M$ give, for some choice of weights $\alpha_1 \ldots \alpha_M$, an integration formula which is exact for (restrictions to the sphere of even) polynomials of degree $\leq 2s$ then the error will be bounded by

$$\sum_{m=0}^{\infty} b_{2m}^2 \left(\frac{1}{C_{2m}} - (A_{2m}^{-1}\mathbb{1}, \mathbb{1})\right)$$

Since the dimension of the space of all homogeneous polynomials of degree $2s$ in (x,y,z) is given by $(s+1)(2s+1)$ it follows that any set of these many directions in "general position" will kill the first $s+1$ terms in the bound for the error.

But if these many points were to be picked carefully we could insure that the tail of the series is very small. This observation points the way to both the possibilities and the dangers of this approach: by picking the directions so as to obtain higher order exact quadrature formulas one could kill some few more terms in the series, but maybe at the expense of making the remaining terms unnecessarily large.

At any rate it should be clear that an exact quadrature formula alone cannot provide a good choice of directions. By the same token one proves that once the desired spatial resolution σ has been fixed -- by fixing

$\phi(\underline{x})$ -- a large enough number of directions in "general position" makes the bound for the error arbitrarily small. This mathematical result can be pretty misleading: all the directions could be confined to a small region in the sphere. Among the good quadrature formulas for the surface of the sphere one should mention those of the product type [23], those found by McLaren [24] by solving systems of linear equations and good to handle polynomials of degree ≤ 14 and those found by Konyaev [25] which can integrate polynomials of degree ≤ 29. The use of these sets of directions, and a number of alternatives has been studied in the forthcoming thesis of A. Tam [26].

One should also mention the observation of A. Louis [27] to the effect that directions which are chosen to avoid the appearance of "low degree ghosts" can give good sets of directions. More explicitly he shows that one can "prune down" some set of directions used by Shepp [28] and get comparable constructions.

Much more work is needed in this area: not only because of the mathematical interest of the problem but mainly for the practical problem of reducing the data collection time with comparable reconstruction quality.

A fair assessment at this time is that product sets as used in [27] or [28] do remarkably well, and are easy to use. These sets do not satisfy any "optimality criterium," but then it is hard to come up with any such criterium for which one can effectively solve for the optimal configuration.

Once the directions have been decided for, the error bound becomes a weighted sum, with positive weights, of the squares of the Gegenbauer coefficients b_{2n} of the function $P\phi$.

Again this points to the possibilities and dangers at hand: one would like to pick $\phi(\underline{x})$ sharply concentrated at the origin to insure good spatial resolution and yet keep all but the first few coefficients b_{2n} pretty small, to provide good contrast resolution.

These two goals are impossible to achieve simultaneously as this situation is covered by the Heisenberg uncertainty principle. Now one knows, see [29], how to obtain the function which comes closest to defeating the Heisenberg principle in the sense that it loses the least (fractional) energy when it is first limited in physical space and then limited in frequency space: one chooses the "first prolate spheroidal wave function".

The choice of $\phi(\underline{x})$ given in [18] was dictated only for practical considerations: one can compute b_{2n} very easily.

These considerations lead to the following

PROBLEM 5. For fixed directions find the "optimal" (not necessarily radial) "desired point response function" $\phi(\underline{x})$. It appears reasonable that in the case of "limited range of views" the best $\phi(\underline{x})$ is not radial.

6. THE LIMITED ANGLE PROBLEM

If the directions ω_j are confined as follows:

$$-\pi/2 < -\Theta \leq \omega_j \leq \Theta < \pi/2$$

in the two dimensional case, or

$$0 \leq \text{colatitude of } \omega_j \leq \Theta < \pi/2$$

in the three dimensional case,

we say that we have a "limited angle problem".

A number of attacks on the reconstruction problem are possible in this case. The list below is by no means complete.

a) Use the same filter for all directions -- say the Shepp-Logan filter-- as if one had unlimited directions.

b) Use the "optimal filters" for the directions ω_j as given in sections 3 and 4 above. This is done in [14, 18].

c) Extrapolate from $Pf(\omega_j)$ to obtain $Pf(\tilde{\omega}_j)$ for a new set of equispaced and unlimited directions $\tilde{\omega}_j$. Then do a standard reconstruction. This method has been used by A. Louis [30].

d) Use iteration techniques in the spirit of Papoulis [31]. This has been done by A. Lent and H. Tuy [32]. The idea is to exploit the fact that if $\hat{g}$ denotes the Fourier transform of g, one has

$$\widehat{Pf(\omega_j)}(\lambda) = \hat{f}(\lambda\omega_j) \qquad \lambda \in R$$

Thus one has to reconstruct a function of <u>compact support</u> in physical space from <u>partial knowledge</u> of its Fourier transform.

A sequence of transforms from physical to Fourier space and back, using at each step the available information to correct the n-th iterate produces (in principle) the function $f(\underline{x})$. A proof can be found in [33], where the author deals with the classical setup of "band and time limiting" and relies heavily on the work of Slepian, Landau and Pollak [29, 34, 35, 36, 37].

This last method is used in electron microscopy among other areas. Great improvements come about by imposing <u>nonlinear corrections</u>, of the type $0 \leq f(\underline{x}) \leq \mathbb{C}$, at each step.

PROBLEM 6. Make a complete analysis of the observed fact that nonlinear corrections render the Papoulis algorithm into a much more robust and stable method.

The use of nonlinear corrections brings in the subject of projecting the unknown function $f(\underline{x})$ onto convex sets instead of ordinary subspaces. For some of these considerations it may be useful to look into E.

Zarantonello's [38] or even better a very recent publication of his available from the University of Pisa, Italy.

All the methods above work well in theory. In practice, noise in the data propagates rather badly: the smaller the value of Θ the more ill-conditioned the problem becomes. A smaller Θ could mean less radiation dosage, less measuring time, etc. but poorer quality of the reconstruction. This suggests the

PROBLEM 7. Find the "expected quality of the reconstruction" as a function of Θ.

This is a poorly defined problem, but yet, an equipment designer may like to know if by reducing Θ in say 10%, the quality in the reconstructions will -- on the average -- go down 5%, 20% or 90%.

A possible way to attack this problem is given in [39]. One studies the singular values λ_i, and two sets of singular vectors $\{f_i\}$ and $\{g_i\}$, [40,41] of the appropriate operator as a function of Θ.

A small singular value λ_i spells troubles in obtaining the component of f along f_i if the data has any component along g_i.

Thus the triple family $\{\lambda_i, f_i, g_i\}$ is needed for a complete analysis. The effort in [39] fails because it is impossible to find these vectors due to numerical limitations.

As B. Marr [42] has pointed out it is better not to go to Fourier space but stay with the Radon transform and thus exploit the findings of Slepian [37]. Elaborating on this, M. Davison [43] has studied the behavior of λ_i as a function of Θ. The same thing can be done for the three dimensional case exploiting this time the analog of Slepian's result which can be found -- among other items -- in [44]. The mathematics of this is treated in [50].

Another attack on this problem is found in [45]. See also [46].

We close with the remark that another good way of picking directions either in the limited angle case if $n = 2$, or in general if $n = 3$, is to take ω_j to be the <u>equilibrium configuration</u> of an array of M electrons attached to the surface of the n-dimensional sphere and subject to whatever extra limitation is present. An explanation of why this is not unexpected is given in [47]. The obvious difficulty here is that of determining explicitly these optimal electrostatic configurations. One can only dream of a machine that would find these ω_j on its own, and use them to do the reconstruction. For numerical results see [26,28]. For experiments on NMR zeugmatography see [48,49].

BIBLIOGRAPHY

1. Marr, R.B., An overview of image reconstruction, Proceedings of the International Symposium on Ill-Posed Problems: Theory and Practice (University of Delaware, Newark, 1979).

2. Hermann, G. editor, Image reconstruction from projections, Springer Verlag, New York, 1979.

3. Radon, J. Über die Bestimmung von Funktionen durch ihre Integral werte längs gewisser Mannifaltigkeiten, Berichte Saechsiche Akademic der Wissenschaften 69, 262-277 (1917).

4. Gelfand, I., Graev, M., and Vilenkin, N., "Generalized functions", vol. 5, English translation, Academic Press (1966).

5. Helgason, S., The Radon transform, Birkhauser, (1980).

6. Uhlenbeck, G., personal communication.

7. Bockwinkel, "On the propagation of light in a two-axial crystal around a center of vibration". Versl. Kon. Akad. v. Wet XIV 2, pp. 636-651, (1906).

8. Uhlenbeck, G., Over Een Stelling Van Lorentz En Haar Uitbreding Voor Meerdimensionale Ruimten (1925).

9. Cramer, H. and Wold, H., London Math. Soc. 11, 290, (1936).

10. Bracewell, R., Strip integration in radio astronomy, Austr. J. Physics 9, 198, (1956).

11. Cormack, A., Representation of a function by its line integrals with some radiological applications, J. Appl. Physics, 34 2722, (1963).

12. Hounsfield, G., British Patent No. 1283915, London, 1972.

13. Smith, K. T., D. C. Solmon and S. L. Wagner, Practical and mathematical aspects of the problem of reconstructiong objects from radiographs, Bulletin of the American Mathematical Society 83 (1977), 1227-1270.

14. Davison, M.E. and F. A. Grünbaum, Convolution algorithms for arbitrary projection angles, IEEE Transactions on Nuclear Science NS-26 (1979), 2670-2673.

15. Cho, Z. H., et al., Computerized image reconstruction methods with multiple photon/X-ray transmission scanning, Phys. Med Biol., 19, 1974, pp. 511-522.

16. Backus, G., and Gilbert, F., Uniqueness in the inversion of inaccurate gross earth data, Phil. Trans. Roy. Soc. Ser. A 266, 1970, pp. 123.

17. Magnus, W., Oberhettinger, F., and Soni, R. P., Formulas and Theorems for the Special Functions of Mathematical Physics, Springer-Verlag, New York, 1966.

18. Davison, M.E. and F. A. Grünbaum, Tomographic reconstruction with arbitrary directions, Communications on Pure and Applied Mathematics 34 (1981), 77-120.

19. Logan, B. F., and Shepp, L. A., Optimal reconstruction of a function from its projections, Duke Math. J., 1975, pp. 645-659.

20. Hamaker, C., and Solmon, D.C., The angles between the null spaces of X-rays, J. Math. Anal. Appl. 62, 1978, pp. 1-23.

21. Maldonado, C.D., Caron, P.A., and Olsen, H.N., New method for obtaining emission coefficients from emmitted spectral intensities. Part II - asymmetrical sources, J. Opt. Soc. Am., 56, 1966, pp. 1305-1313.

22. Miller, K., An optimal method for the X-ray reconstruction problem, preliminary report, Amer. Math. Soc. Notices January, 1978, A-161.

23. Stroud, A.H., Approximate calculation of multiple integral, Prentice Hall, 1971.

24. McLaren, A.D., Optimal numerical integration on a sphere, Math. Comput. V. 17, 1963, pp. 361-383.

25. Konyaev, S. I., (1979) Quadratures of Guassian type for a sphere invariant under the icosahedral group with inversion, Math. Note, Vol. 25, No. 4, 629-634, April 1979.

26. Tam, A., Thesis, University of California at Berkeley, 1982.

27. Louis, A., Optimal sampling in nuclear magnetic resonance zeugmatography, (1981).

28. Shepp, L. A., Computerized tomography and nuclear magnetic resonance, Journal of Computer Assisted Tomography, 4 (1980), 94-107.

29. Slepian, D. and Pollak, H. O., "Prolate spheroidal wave functions, Fourier Analysis and Uncertainty I," Bell System Tech. Journal, 40, No. 1, (1961) 43-64.

30. Louis, A., Picture reconstruction from projections in restricted range, Mathematical Methods in the Applied Sciences, 2 (1980), 209-220.

31. Papoulis, A., A new algorithm in spectral analysis and band-limited extrapolation, IEEE Trans. Cir. and Syst., vol. CAS-22, No. 9, Spet. 1975.

32. Lent, A., and Tuy, H., An Iterative Method for the Extrapolation of Band-Limited Functions, Buffalo, 1979.

33. Youla, D.C., Generalized image restoration by the method of alternating orthogonal projections, IEEE Trans. on Circuits and Systems, vol. CAS-25, No. 9 (1978).

34. Landau, H.J. and Pollak, H.O., "Prolate spheroidal wave functions, Fourier Analysis and Uncertainty II," Bell System Tech. Journal, 40, No. 1, (1961), 65-84.

35. Landau, H.J. and Pollak, H.O., "Prolate spheroidal wave functions, Fourier Analysis and Uncertainty III," Bell System Tech. Journal, 41, No. 4, (1962), 1295-1336.

36. Slepian, D., "Prolate spheroidal wave functions, Fourier Analysis and Uncertainty IV," Bell System Tech. Journal, 43, No. 6 (1964) 3009-3058.

37. Slepian, D., "Prolate spheroidal wave functions, Fourier Analysis and Uncertainty V," Bell System Tech. Journal, 57, No. 5 (1978) 1371-1430.

38. Zarantonello, E. Projections on Convex Sets in Hilbert Space and spectral theory, Academic Press, 1971.

39. Grünbaum, F.A., A study of Fourier space methods for "limited angle" image reconstruction, Numerical Functional Analysis and Optimization 2 (1), (1980), 31-42.

40. Noble, B., Applied Linar Algebra, Prentice-Hall, Englewood Cliffs, N.J., 1969.

41. Nashed, Zuhair M., ed., Generalized Inverses and Applications, Academic Press, 1976, p. 193, et seq.

42. Marr, B., private communication.

43. Davison, M.E., The ill conditioned nature of the limited angle tomography problem, to appear in Siam J. Applied Math.

44. Grünbaum, F.A., Longhi, L., and Perlstadt, M., Differential operators commuting with finite convolution integral operators: Some nonabelian examples, SIAM Journal on Applied Mathematics (1982).

45. Logan, B., The Uncertainty principle in reconstructing functions from projections, Duke Math. J. 42, 4, 1975.

46. Davison, M.E. A singular value decomposition for the Radon transform in n-dimensional euclidean space, Numerical Funct. Anal. and Optimization 3(3) 321-340. (1981).

47. F. A. Grünbaum, Reconstruction with arbitrary directions: dimensions two and three, in Mathematical Aspects of Computerized Tomography, Lecture Notes in Medical Informatics, Vol. 8, G.T. Herman and F. Natterer, eds., Springer-Verlag, 1981, pp. 112-126.

48. P.C. Lauterbur (1973). Image formation by induced local interactions: examples employing nuclear magnetic resonance. Nature (London) 242, 190-191.

49. R.B. Marr, C. Chen and P.C. Lauterbur, On two approaches to 3D reconstruction in NMR zeugmatography, in Mathematical Aspects of Computerized Tomography, Lecture Notes in Medical Informatics, Vol. 8, G.T. Herman and F. Natterer, eds., Springer Verlag, 1981.

50. F.A. Grünbaum, The Slepian-Landau-Pollak phenomenon for the radial part of the Laplacian (1981).

DEPARTMENT OF MATHEMATICS
UNIVERSITY OF CALIFORNIA
BERKELEY, CALIFORNIA 94720

Proceedings of Symposia in Applied Mathematics
Volume 27, 1982

RANGES OF RADON TRANSFORMS

Sigurdur Helgason

1. Introduction.

Let X be a differentiable manifold in which certain submanifolds ξ depending smoothly on parameters $(\xi_1,\ldots,\xi_n)$ are singled out. Given a function f on X its Radon transform $\hat{f}$ is defined by

$$\hat{f}(\xi) = \hat{f}(\xi_1,\ldots,\xi_n) = \int_\xi f(x)\, d\sigma(x),$$

$d\sigma$ being some specified measure on ξ. Let Ξ denote the set $\{\xi\}$ of manifolds ξ.

A central problem is to invert (1), that is recover f from its integrals over the manifolds ξ. Another problem of significance for applications of the Radon transform is the determination of the space of functions $\hat{f}$ on Ξ as f runs through some naturally defined function space F on X. In this lecture we shall focus attention on this latter problem and discuss a sample of old and new results for it. The corresponding inversion formulas, although known for most of the cases, will be left out.

The most natural spaces to consider for F are the space $\mathcal{E}(X) = C^\infty(X)$ of C^∞-functions on X and the space $\mathcal{D}(X) = C_c^\infty(X)$ of C^∞ functions on X of compact support. Extending (1) in a natural way to distributions we shall also discuss the image under the Radon transform of the space $\mathcal{E}'(X)$ of distributions on X of compact support.

Naturally one cannot expect a unified description of the space $\mathcal{D}(X)^\wedge$; however, it is very interesting to compare the answers for different cases of X.

2. Spheres and Two-Point Homogeneous Spaces.

Let $X = \mathbf{S}^2$ and let Ξ denote the set of great circles in $\mathbf{S}^2$. Let F denote the space of even C^∞ functions on $\mathbf{S}^2$. In a paper [4], preceding Radon's paper [8], Funk proved that an even function on $\mathbf{S}^2$ is determined by its integrals over great circles. In other words, the mapping $f \longrightarrow \hat{f}$ is

Presented at the A.M.S. Short Course on Computed Tomography, Jan. 12, 1982.

0160-7634/82/0000-0878/$03.00

one-to-one on F.

We can view F as the space $\mathcal{E}(\mathbb{P}^2(\mathbb{R}))$ of C^∞ functions on the elliptic space $\mathbb{P}^2(\mathbb{R}) = \mathbf{S}^2/\mathbb{Z}_2$, which is obtained from the sphere by identification of antipodal points. But then Ξ can be identified with $\mathbb{P}^2(\mathbb{R})$ so the mapping $f \longrightarrow \hat{f}$ is a one-to-one mapping of F into F. Using spherical harmonics one can prove that this mapping is also surjective.

These results for $\mathbf{S}^2$ generalize to an arbitrary compact two-point homogeneous space X as follows. Given $x \in X$, let A_x denote the corresponding antipodal manifold, that is the set of points in X at maximum distance from x. Let Ξ denote the set of antipodal manifolds. The mapping $j: x \longrightarrow A_x$ being one-to-one, Ξ acquires a differentiable structure by requiring j to be a diffeomorphism. We then have the following result from [5(b)].

THEOREM 2.1. *Let X be a compact two-point homogeneous space. Then the Radon transform $f \longrightarrow \hat{f}$ is a bijection of $\mathcal{E}(X)$ onto $\mathcal{E}(\Xi)$.*

3. Points and Hyperplanes in $\mathbb{R}^n$.

Here we let $X = \mathbb{R}^n$ and let Ξ denote the manifold of hyperplanes in X. Writing a hyperplane ξ in the form $(x,\omega) = p$ where $\omega = (\omega_1,\ldots,\omega_n) \in \mathbb{R}^n$ is a unit vector, $(\ ,\ )$ the inner product, and $p \in \mathbb{R}$, we write (1) in the form

$$(2) \qquad \hat{f}(\omega,p) = \int_{(x,\omega)=p} f(x)\,dm(x), \quad \hat{f}(\omega,p) = \hat{f}(-\omega,-p),$$

dm being the Euclidean measure. The dual Radon transform $\phi \longrightarrow \check{\phi}$ associates to a continuous function ϕ on Ξ the function

$$(3) \qquad \check{\phi}(x) = \int_{x\in\xi} \phi(\xi)\,d\mu(\xi),$$

where μ is the normalized measure on the compact set $\{\xi \in \Xi : x \in \xi\}$ which is invariant under the rotations around x. If $S \in \mathcal{E}'(X)$ the Radon transform $\hat{S} \in \mathcal{E}'(\Xi)$ is defined by (cf. [5(c), 5(g)]),

$$(4) \qquad \hat{S}(\phi) = S(\check{\phi}), \qquad \phi \in \mathcal{E}(\Xi).$$

Also if Σ is a distribution on Ξ its dual transform is defined by

$$(5) \qquad \check{\Sigma}(f) = \Sigma(\hat{f}), \qquad f \in \mathcal{D}(X).$$

The following characterization of $\mathcal{D}(X)^\wedge$ was given in [5(a)], Theorem 3.2.

THEOREM 3.1. *The range $\mathcal{D}(\mathbb{R}^n)^\wedge$ consists of the functions $\phi(\omega,p) \in \mathcal{D}(\Xi)$ satisfying the condition:*

For each $k \in \mathbb{Z}^+$ the function

$$(6) \qquad \int_{\mathbb{R}} \phi(\omega,p)p^k dp$$

is a homogeneous polynomial in $\omega_1,\dots,\omega_n$ *of degree* k.

This theorem can be put into another form which makes it very easy to write down specific functions in the range $\mathcal{D}(\mathbb{R}^n)$. For simplicity in notation we take $n = 2$.

THEOREM 3.2. *The range* $\mathcal{D}(\mathbb{R}^2)^\wedge$ *consists of the functions* $\phi \in \mathcal{D}(\Xi)$ *which when expanded in Fourier series*

$$(7) \qquad \phi(e^{i\theta},p) = \sum_{n\in\mathbb{Z}} \phi_n(p)e^{in\theta}$$

have the property: For each $n \in \mathbb{Z}$

$$(8) \qquad \phi_n(p) = \frac{d^{|n|}}{dp^{|n|}} h_n(p)$$

where $h_n \in \mathcal{D}(\mathbb{R})$ *is even.*

These results can be generalized to distributions.

THEOREM 3.3. *The range* $\mathcal{E}'(\mathbb{R}^n)^\wedge$ *consists of the distributions* $\Sigma \in \mathcal{E}'(\Xi)$ *with the property:*

For each $k \in \mathbb{Z}^+$ *the distribution*

$$(9) \qquad \phi \in C^\infty(\mathbf{S}^{n-1}) \longrightarrow \int_{\mathbf{S}^{n-1}\times\mathbb{R}} \phi(\omega)p^k d\Sigma(\omega,p)$$

is a homogeneous polynomial in $\omega_1,\dots,\omega_n$ *of degree* k.

We sketch the proof. It is rather easy to prove, by approximation, that each Σ in the image $(\mathcal{E}')^\wedge$ has the property stated. For the converse suppose Σ has property (9). If the operator $\Lambda\colon \mathcal{D}(\Xi) \longrightarrow \mathcal{E}(\Xi)$ is defined by

$$(\Lambda\phi)(\omega,p) = \begin{cases} \dfrac{d^{n-1}}{dp^{n-1}}\phi(\omega,p) & n \text{ odd} \\[2ex] \mathcal{H}_p \dfrac{d^{n-1}}{dp^{n-1}}\phi(\omega,p) & n \text{ even} \end{cases}$$

($\mathcal{H}$ = Hilbert transform) then we define a distribution S on X by

$$(10) \qquad S(f) = \Sigma(\Lambda\hat{f}), \qquad f \in \mathcal{D}(X).$$

This is motivated by the formula ([7], §1 or [5(g)], §§ 4-5)

$$\int_X f(x)g(x)dx = \int_\Xi (\Lambda\hat{f})(\xi)\hat{g}(\xi)d\xi, \qquad f,g \in \mathcal{D}(X),$$

where $d\xi$ is a suitable constant multiple of the invariant measure $d\omega\, dp$ on Ξ. Let G be the group of isometries of X and dg a Haar measure on it. If $g \in G$ let S^g denote the image of S under g (i.e. $S^g(f)=\int f(g\bullet x)dS(x)$) and define Σ^g similarly. If $F \in \mathcal{D}(G)$ the distribution

$$f \longrightarrow \int_G F(g) S^g(f)\,dg, \qquad f \in \mathcal{D}(X)$$

is given by a compactly supported function s on X. Similarly, the distribution

$$\phi \longrightarrow \int_G F(g) \Sigma^g(\phi)\,dg$$

is given by a compactly supported function σ on Ξ. Using Theorem 3.1 and the inversion formula for the Radon transform one concludes that $\hat{s} = c\sigma$ where c is a constant. Then a suitable limiting argument gives the desired conclusion $\hat{S} = c\Sigma$.

Remark 1. This theorem can also be reformulated in the style of Theorem 3.2 (cf. Theorem 4.2 below).

Remark 2. In an unpublished manuscript, Ambrose has defined a Radon transform on $\mathcal{E}'(\mathbb{R}^n)$ with a description of the range. However, it is not clear what relation his definition has to ours. In [7], §4, the Radon transform of all distributions is considered but we have found the treatment rather uncertain.

4. The Hyperbolic Plane.

Next we consider the case when X is the hyperbolic plane, that is the open unit disk $x_1^2 + x_2^2 < 1$ with the Riemannian metric

$$ds^2 = \frac{dx_1^2 + dx_2^2}{(1 - x_1^2 - x_2^2)^2} .$$

For the manifolds ξ we take the horocycles, that is the circles tangential to the boundary $x_1^2 + x_2^2 = 1$. Since the geodesics in X are the arcs perpendicular to the boundary the horocycles are orthogonal trajectories to parallel geodesics. Thus they are analogs to hyperplanes which are indeed orthogonal surfaces to parallel lines. Let $\xi(\omega,t)$ denote the horocycle tangential to the unit circle at ω and having (Riemannian) distance t from the origin 0 (see Fig. 1). The distance t is taken with sign (negative if 0 is inside the horocycle). The parametrization $(\omega,t) \longrightarrow \xi(\omega,t)$ identifies the set Ξ with the cylinder $\mathbf{S}' \times \mathbb{R}$. The Radon transform can be written

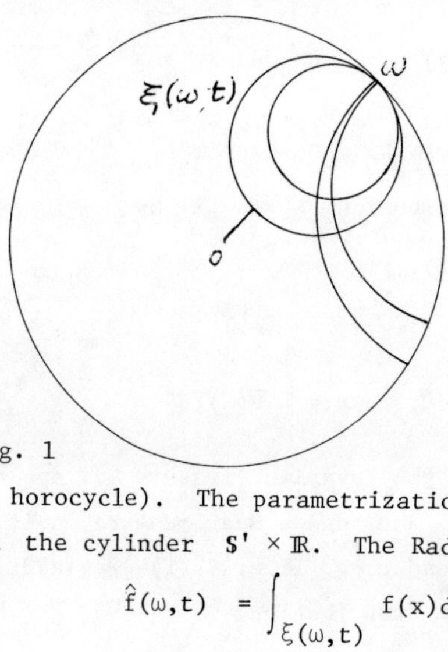

Fig. 1

$$\hat{f}(\omega,t) = \int_{\xi(\omega,t)} f(x)\,d\sigma_t(x), \tag{11}$$

where $d\sigma_t$ is the (Riemannian) arc length on $\xi(\omega,t)$. Our problem is to characterize the functions $\hat{f}(\omega,t)$ as f runs through $\mathcal{D}(X)$. One sees quickly that an arbitrary function in $\mathcal{D}(\Xi)$ does not arise as a function $\hat{f}(\omega,t)$; in fact if we decompose X into parallel horocycles (tangential at ω) the Riemannian surface element dx on X decomposes accordingly,

$$\int_X f(x)dx = \int_{\mathbb{R}} dt \int_{\xi(\omega,t)} f(x)d\sigma(x) = \int_{\mathbb{R}} \hat{f}(\omega,t)dt \tag{12}$$

so

$$\int_{\mathbb{R}} \hat{f}(\omega,t)dt \text{ is independent of } \omega. \tag{13}$$

A characterization of the space $\mathcal{D}(X)^\wedge$ is given in [5(d)], Theorem 8.4 for the more general case when X is a symmetric space of the noncompact type. However, this characterization is rather unwieldy and we shall now write down a more effective formulation for the case at hand. This formulation generalizes to symmetric spaces X as well, see remark below. For some support theorems for L^2 functions on the three-dimensional hyperbolic space see Lax-Phillips [6].

Writing $\omega = e^{i\theta}$ a function $\phi \in \mathcal{D}(\Xi)$ can be expanded in a Fourier series in the ω-variable,

$$\phi(\xi(e^{i\theta},t)) = \sum_{n\in\mathbf{Z}} \phi_n(t)e^{in\theta},$$

where

$$\phi_n(t) = \frac{1}{2\pi}\int_0^{2\pi} \phi(\xi(e^{i\theta},t))e^{-in\theta}d\theta.$$

THEOREM 4.1. The horocycle Radon transform is a one-to-one mapping of $\mathcal{D}(X)$ onto the space of functions $\psi \in \mathcal{D}(\Xi)$ which when expanded

$$\psi(\xi(e^{i\theta},t)) = \sum_{n\in\mathbb{Z}} \psi_n(t)e^{in\theta} \tag{15}$$

satisfy

$$\psi_n(t) = e^{-t}\left(\frac{d}{dt}-1\right)\left(\frac{d}{dt}-3\right)\cdots\left(\frac{d}{dt}-2|n|+1\right)\phi_n(t), \tag{16}$$

where $\phi_n \in \mathcal{D}(\mathbb{R})$ is even.

This description makes it easy to write down specific functions in the range $\mathcal{D}(X)^\wedge$. Note also that since

$$e^{-t}\left(\frac{d}{dt}-1\right)(e^t\phi) = \frac{d\phi}{dt}$$

the function $\psi_n(t)$ is, for $|n| > 0$ the derivative of a compactly supported function so $\int\psi_n = 0$. In particular,

$$\int_{\mathbb{R}} \psi(\xi(e^{i\theta},t))dt \text{ is independent of } \theta \text{ in accordance}$$

with (12).

Remark. The space X has constant curvature -4. For curvature $-\varepsilon$ the condition (16) holds with right hand side replaced by

$$e^{-k_0 t}\left(\frac{d}{dt} - k_1\right) \cdots\cdots \left(\frac{d}{dt} - k_{|n|}\right)\phi_n(t),$$

where $k_i = O(\varepsilon)$. Thus Theorem 3.2 appears as a limiting case as $\varepsilon \longrightarrow 0$.

Theorem 4.1 generalizes to the space $\mathcal{E}'(X)$ of distributions of compact support. In order to describe this we define for each $\sigma \in \mathcal{D}'(\Xi)$ and each $n \in \mathbb{Z}$ the distribution $\sigma_n \in \mathcal{D}'(\mathbb{R})$ by

$$\sigma_n(\phi) = \sigma(e^{-2t}\phi(t) \otimes e^{in\theta}). \tag{17}$$

Then, in the notation of (15) we have

$$\sigma(\psi) = \sum_n (e^{2t}\sigma_n)(\psi_{-n}).$$

The reason for the factor e^{2t} is that in the parametrization $(\theta,t) \longrightarrow \xi(e^{i\theta},t)$ of Ξ the measure on Ξ invariant under all isometries of X is given by $e^{2t}dt\, d\theta$. We then have the following extension of Theorem 4.1 to distributions.

THEOREM 4.2. *The horocycle Radon transform is a bijection of* $\mathcal{E}'(X)$ *onto the set of distributions* $\sigma \in \mathcal{E}'(\Xi)$ *whose Fourier coefficients* σ_n *as defined by* (17) *have the form*

$$\sigma_n = e^{-t}\left(\frac{d}{dt} - 1\right)\cdots\left(\frac{d}{dt} - 2|n|+1\right)\tau_n,$$

where $\tau_n \in \mathcal{D}'(\mathbb{R})$ *is even*.

Remark. Using the methods of [5(e)], Theorems 4.1-4.2 extend to symmetric spaces of the noncompact type. The differential operators in (16) are then replaced by certain matrices of partial differential operators. Since the required notation is rather elaborate we postpone this to another occasion.

For symmetric spaces of rank one, in particular the higher dimensional hyperbolic spaces, (16) remains true with a minor change if (15) is replaced by the expansion of $\psi(\xi(\omega,t))$ in spherical harmonics.

5. The X-ray Transform.

Let X be a complete Riemannian manifold of dimension > 1 in which any two points can be joined by a unique geodesic. Let Ξ denote the set of all (complete) geodesics ξ in X. The *X-ray transform* on X is the mapping $f \longrightarrow \hat{f}$ where

$$\hat{f}(\xi) = \int_\xi f(x)\,ds(x),$$

ds denoting the element of arc length. If d denotes the distance in X and 0 some fixed point we let F denote the set of continuous functions f on X

satisfying

$$\sup_x e^{kd(0,x)}|f(x)| < \infty \quad \text{for each} \quad k \geq 0.$$

Here we have the following support theorem (or "hole theorem"), cf.[5(f)].

THEOREM 5.1. *Let* X *be the hyperbolic plane or, more generally, a symmetric space of the noncompact type. Let* B *be any closed ball in* X.

Suppose $f \in F$ *satisfies*

$$\hat{f}(\xi) = 0 \quad \text{whenever} \quad \xi \quad \text{is outside} \quad B.$$

Then

$$f(x) = 0 \quad \text{for} \quad x \quad \text{outside} \quad B.$$

As a consequence of Theorem 3.1, the image $\mathcal{D}(\mathbb{R}^n)^\wedge$ of $\mathcal{D}(\mathbb{R}^n)$ under the X-ray transform can be explicitly described (cf. [5(g)], Cor. 6.4; an L^2 version is given in [10], §6). The analog of this result for the X-ray transform on the hyperbolic space has however not been proved. The difficulty is already clear from the lack of an analog to (13). In fact, if we decompose the hyperbolic plane into geodesics perpendicular to the x-axis as in Fig.2 we have

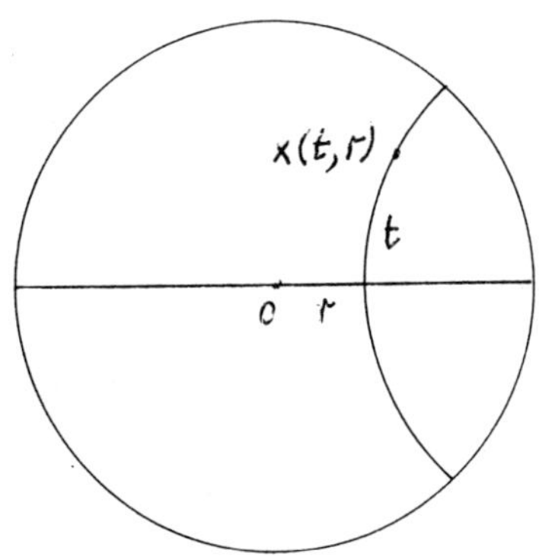

Fig. 2

$$\int_X f(x)dx = \int_{\mathbb{R}}\int_{\mathbb{R}} f(x(t,r))\cosh 2t \, dt \, dr$$

in contrast to (12) so (13) does not have an analog here.

6. Spheres through the Origin.

Our final example is $X = \mathbb{R}^n$ and Ξ the set of all spheres passing through the origin 0. Let as usual $S_r(x_0)$ denote the sphere $|x - x_0| = r$ in $\mathbb{R}^n$. The transform $f \longrightarrow \hat{f}$ is now given by

$$\hat{f}(2y) = \int_{S_{|y|}(y)} f(x)d\omega(x), \quad y \in \mathbb{R}^n,$$

where $d\omega$ is the Euclidean surface element. For this transform which has been studied by Cormack [2], Chen [1] and Rhee [9], the following result about the range has been obtained by Cormack-Quinto [3].

THEOREM 6.1. *Let F denote the set of restrictions to $\mathbb{R}^n$ of the holomorphic functions on $\mathbb{C}^n$. Then $f \longrightarrow \hat{f}$ is a bijection of F onto itself.*

BIBLIOGRAPHY

1. Chen, Y.W. "On the solutions of the wave equation in a quadrant of $\mathbb{R}^4$." Bull. Amer. Math. Soc. 70(1964), 172-177.

2. Cormack A. "Representation of a function by its line integrals, with some radiological applications I, II", J. Appl. Physics, 34, (1963), 2722-2727; 35(1964), 2908-2912.

3. Cormack, A. and E.T. Quinto, "A Radon transform on spheres through the origin in $\mathbb{R}^n$ and applications to the Darboux equation", Trans. Amer. Math. Soc.,260(1980), 575-581.

4. Funk, P. "Über eine geometrische Anwendung der Abelschen Integralgleichung", Math. Ann. 77(1916), 129-135.

5. Helgason, S. (a) "A duality in integral geometry; some generalizations of the Radon transform", Bull. Amer. Math. Soc. 70(1964), 435-446.

____________ (b) "The Radon transform on Euclidean spaces, compact two-point homogeneous spaces and Grassmann manifolds", Acta Math. 113(1965), 153-180.

____________ (c) "A duality in integral geometry on symmetric spaces", Proc. U.S.-Japan Seminar in Differential Geometry, Kyoto, 1965. Nippon Hyoransha, Tokyo 1966, 37-56.

____________ (d) "The surjectivity of invariant differential operators on symmetric spaces", I. Ann. of Math. 98(1973), 451-479.

____________ (e) "A duality for symmetric spaces with applications to group representations II. Differential equations and eigenspace representations", Advances in Math. 22(1976), 187-219.

____________ (f) "The X-ray transform on a symmetric space", Proc. Conf. Diff. Geom. and Global Analysis, Berlin 1979, Springer Lecture Notes, No. 836, 1981.

____________ (g) "The Radon Transform", Birkhäuser, Boston, 1980.

6. Lax, P.D. and R.S. Phillips, "Translation representation for the solutions of the non-Euclidean wave equation", Comm. Pure Appl. Math. 32(1979), 617-667.

7. Ludwig, D. "The Radon transform on Euclidean space", Comm. Pure Appl. Math. 23(1966), 49-81.

8. Radon, J. "Über die Bestimmung von Funktionen durch ihre Integralwerte längs gewisser Mannigfaltigkeiten", Ber. Verh. Sächs. Akad. Wiss. Leipzig Math. Nat. kd. 69(1917), 262-277.

9. Rhee, H. "A representation of the solutions of the Darboux equation in odd-dimensional spaces", Trans. Amer. Math. Soc. 150(1970), 491-498.

10. Solmon, D.C. "The X-ray transform", J. Math. Anal. Appl. 56(1976), 61-83.

Proceedings of Symposia in Applied Mathematics
Volume 27, 1982

SITZUNG VOM 30. APRIL 1917.

Über die Bestimmung von Funktionen durch ihre Integralwerte längs gewisser Mannigfaltigkeiten.

Von

JOHANN RADON.

Integriert man eine geeigneten Regularitätsbedingungen unterworfene Funktion zweier Veränderlichen x, y — eine *Punktfunktion* $f(P)$ in der Ebene — längs einer beliebigen Geraden g, so erhält man in den Integralwerten $F(g)$ eine *Geradenfunktion.* Das in Abschnitt A vorliegender Abhandlung gelöste Problem ist die Umkehrung dieser linearen Funktionaltransformation, d. h. es werden folgende Fragen beantwortet: kann jede, geeigneten Regularitätsbedingungen genügende Geradenfunktion auf diese Weise entstanden gedacht werden? Wenn ja, ist dann f durch F eindeutig bestimmt und wie kann es ermittelt werden?

Im Abschnitte B gelangt das dazu in gewisser Hinsicht duale Problem der Bestimmung einer Geradenfunktion $F(g)$ aus ihren Punktmittelwerten $f(P)$ zur Lösung.

Schließlich werden im Abschnitte C gewisse Verallgemeinerungen kurz besprochen, zu denen insbesondere die Betrachtung nichteuklidischer Mannigfaltigkeiten sowie höherer Räume Anlaß gibt.

Die Behandlung dieser an sich interessanten Probleme gewinnt ein erhöhtes Interesse durch die zahlreichen Beziehungen, die zwischen diesem Gegenstande und der Theorie des logarithmischen und NEWTONschen Potentiales bestehen, auf die an den bezüglichen Stellen zu verweisen sein wird.

Reprinted by permission of the BSB B. G. Teubner Verlagsgesellschaft, from *Berichte über die Verhandlungen der Königlich Sächsischen Gesellschaft der Wissenschaften zu Leipzig* (*Leipzig,* 1917), pp. 262–277.

A. Bestimmung einer Punktfunktion in der Ebene aus ihren geradlinigen Integralwerten.

1. Es sei $f(x, y)$ eine für alle reellen Punkte $P = [x, y]$ erklärte reelle Funktion, die folgende Regularitätsbedingungen erfülle:

a_1) $f(x, y)$ sei stetig.

b_1) Es konvergiere das über die ganze Ebene zu erstreckende Doppelintegral

$$\iint \frac{|f(x, y)|}{\sqrt{x^2 + y^2}} dx\, dy.$$

c_1) Wird für einen beliebigen Punkt $P = [x, y]$ und jedes $r \geqq 0$

$$\bar{f}_P(r) = \frac{1}{2\pi} \int_0^{2\pi} f(x + r\cos\varphi,\ y + r\sin\varphi)\, d\varphi$$

gesetzt, so gelte für jeden Punkt P:

$$\lim_{r \to \infty} \bar{f}_P(r) = 0.$$

Dann gelten folgende Sätze:

Satz I: Der geradlinige Integralwert von f für die Gerade g mit der Gleichung $x\cos\varphi + y\sin\varphi = p$, der durch

$$\text{(I)} \quad F(p, \varphi) = F(-p, \varphi + \pi) = \int_{-\infty}^{+\infty} f(p\cos\varphi - s\sin\varphi,\ p\sin\varphi + s\cos\varphi)\, ds$$

gegeben ist, ist „im allgemeinen" vorhanden; das soll heißen: auf jedem Kreise bilden die Berührungspunkte jener Tangenten, für welche F nicht existiert, eine Menge vom linearen Maße Null:

Satz II: Bildet man den Mittelwert von $F(p, \varphi)$ für die Tangenten des Kreises mit dem Zentrum $P = [x, y]$ und dem Radius q:

$$\text{(II)} \qquad \bar{F}_P(q) = \frac{1}{2\pi} \int_0^{2\pi} F(x\cos\varphi + y\sin\varphi + q,\ \varphi)\, d\varphi,$$

so konvergiert dieses Integral für alle P, q absolut.

Satz III: Der Wert von f ist durch F eindeutig bestimmt und läßt sich folgendermaßen berechnen:

$$\text{(III)} \qquad f(P) = -\frac{1}{\pi} \int_0^{\infty} \frac{d\bar{F}_P(q)}{q}.$$

Dabei ist das Integral im STIELTJES*schen Sinne zu verstehen und kann auch durch die Formel:*

$$(\text{III}') \qquad f(P) = \frac{1}{\pi} \lim_{\varepsilon \to 0} \left(\frac{\overline{F}_P(\varepsilon)}{\varepsilon} - \int_{\varepsilon}^{\infty} \frac{\overline{F}_P(q)}{q^2} dq \right)$$

definiert werden.

Indem wir zum Beweise schreiten, bemerken wir vorweg, daß die Bedingungen $a_1 - c_1$ gegenüber Bewegungen der Ebene invariant sind. Wir können also den Punkt [0, 0] immer als Repräsentanten irgendeines Punktes der Ebene betrachten.

Man erkennt nun das Doppelintegral:

$$(1) \qquad \iint\limits_{x^2+y^2>q^2} \frac{f(x, y)}{\sqrt{x^2+y^2-q^2}} dx\, dy$$

als absolut konvergent. Vermöge der Transformation

$$x = q \cos \varphi - s \sin \varphi, \quad y = q \sin \varphi + s \cos \varphi$$

geht dasselbe über in:

$$\int_0^{2\pi} d\varphi \int_0^{\infty} f(q \cos \varphi - s \sin \varphi,\ q \sin \varphi + s \cos \varphi) ds$$

$$= \int_0^{2\pi} d\varphi \int_{-\infty}^{0} f(q \cos \varphi - s \sin \varphi,\ q \sin \varphi + s \cos \varphi) ds,$$

so daß man seinen Wert auch durch

$$\frac{1}{2} \int_0^{2\pi} d\varphi \int_{-\infty}^{+\infty} f(q \cos \varphi - s \sin \varphi,\ q \sin \varphi + s \cos \varphi) ds = \frac{1}{2} \int_0^{2\pi} F(q, \varphi) d\varphi$$

ausdrücken kann. Nach bekannten Eigenschaften der absolut konvergenten Doppelintegrale folgen hieraus die Behauptungen der Sätze I und II.

Um auf die Formel (III) zu kommen, kann man folgenden Weg einschlagen: Einführung von Polarkoordinaten in (1) liefert

$$\int_q^{\infty} dr \int_0^{2\pi} \frac{f(r \cos \varphi,\ r \sin \varphi)}{\sqrt{r^2 - q^2}} d\varphi$$

oder mit Hilfe der Mittelwertsbezeichnung von c_1:

$$2\pi\int_q^\infty \frac{\bar{f}_0(r)\,dr}{\sqrt{r^2-q^2}}.$$

Verglichen mit dem zuletzt erhaltenen Werte von (1) resultiert:

$$\bar{F}_0(q) = 2\int_q^\infty \frac{\bar{f}_0(r)\,dr}{\sqrt{r^2-q^2}}. \tag{2}$$

Führt man in dieser Integralgleichung erster Art die Variablen $r^2 = v$, $q^2 = u$ ein, so kann man sie leicht nach Art der bekannten Abelschen lösen und erhält die Formel (III) für

$$\bar{f}_0(0) = f(0, 0).$$

Bei dieser Ableitung erscheint es aber schwer, ohne weitere Bedingungen für f auszukommen, daher geben wir einer direkten Verifikation den Vorzug.

Zunächst ist, um die Gleichheit der Ausdrücke (III) und (III') zu zeigen, zu beweisen, daß:

$$\lim_{q\to\infty} \frac{\bar{F}_0(q)}{q} = 0.$$

Vermöge (2) ist

$$\left|\frac{\bar{F}_0(q)}{q}\right| \leqq \frac{2}{q}\left|\int_q^{2q} \frac{\bar{f}_0(r)}{\sqrt{r^2-q^2}}\right| + \frac{2}{q}\left|\int_{2q}^\infty \frac{\bar{f}_0(r)\,r\,dr}{\sqrt{r^2-q^2}}\right|$$

$$\leqq 2\sqrt{3}\,|\bar{f}_0(t)| + \frac{4}{\sqrt{3}}\int_{2q}^\infty |\bar{f}_0(r)|\,dr \qquad (q<t<2q)$$

und dies konvergiert wegen b_1 und c_1 für $q\to\infty$ gegen Null.

Die rechte Seite von (III') geht nun durch Einführung von (2) über in:

$$\frac{2}{\pi}\lim_{\varepsilon\to 0}\left[\frac{1}{\varepsilon}\int_\varepsilon^\infty \frac{r\,\bar{f}_0(r)}{\sqrt{r^2-\varepsilon^2}}dr - \int_\varepsilon^\infty \frac{dq}{q^2}\int_q^\infty \frac{r\,\bar{f}_0(r)}{\sqrt{r^2-q^2}}dr\right].$$

Vertauscht man in dem zweiten Integral die Integrationsfolge, so kann man nach q integrieren, erkennt dabei das Integral als absolut konvergentes Doppelintegral, was die Vertauschung rechtfertigt, und findet für obigen Ausdruck den Wert

$$\frac{2}{\pi} \lim_{\varepsilon \to 0} \int_{\varepsilon}^{\infty} \frac{\bar{f}_0(r)}{r\sqrt{r^2 - \varepsilon^2}} dr,$$

was tatsächlich $\bar{f}_0(0) = f(0, 0)$ liefert, wie unschwer zu zeigen ist.

2. Es sei $F(p, \varphi) = F(-p, \varphi + \pi)$ eine Geradenfunktion, die folgende Regularitätsbedingungen erfüllt:

a_2) F und die Ableitungen F_p, F_{pp}, F_{ppp}, F_φ, $F_{p\varphi}$, $F_{pp\varphi}$ seien für alle $[p, \varphi]$ stetig.

b_2) F, F_φ, pF_p, $pF_{p\varphi}$ und pF_{pp} konvergieren für $p \to \infty$ gleichmäßig in φ gegen Null.

c_2) Die Integrale:

$$\int_0^\infty F_{pp}\, lp\, dp, \quad \int_0^\infty F_{ppp}\, plp\, dp, \quad \int_0^\infty F_{pp\varphi}\, plp\, dp$$

konvergieren absolut und gleichmäßig in φ.

Dann können wir beweisen:

Satz IV: Bildet man nach (III) bzw. (III′) $f(P)$, so erfüllt dieses die Bedingungen a_1, b_1, c_1 und liefert als geradlinige Integralwerte das vorgelegte $F(p, \varphi)$. Infolge Satz III ist es die einzige derartige Funktion.

Führen wir in (III) Polarkoordinaten ein, so ergibt sich:

$$f(\varrho \cos\psi, \varrho \sin\psi) = -\frac{1}{2\pi^2}\int_0^\infty \frac{dp}{p}\int_0^{2\pi} F_p(\varrho\cos\omega + p, \omega + \psi)\, d\omega$$

$$= \frac{1}{2\pi^2}\int_0^\infty lp\, dp \int_0^{2\pi} F_{pp}(p + \varrho\cos\omega, \omega + \psi)\, d\omega.$$

Es ist nämlich:

$$\int_0^{2\pi} F_p(\varrho\cos\omega + p, \omega + \psi)\, d\omega = \int_0^{2\pi} F_p(\varrho\cos\omega, \omega + \psi)\, d\omega$$

$$+ \int_0^{2\pi} d\omega \int_0^p F_{pp}(\varrho\cos\omega + t, \omega + \psi)\, dt$$

und der erste Bestandteil ist wegen $F(p, \varphi) = F(-p, \varphi + \pi)$ gleich Null. Deshalb konvergiert das Produkt des Integrals mit lp

für $p \to 0$ gegen Null. Auf Grund derselben Eigenschaft von F folgt weiter:

$$(3)\quad f(\varrho\cos\psi, \varrho\sin\psi) = \frac{1}{2\pi^2}\int_0^{\pi} d\omega \int_{-\infty}^{+\infty} F_{pp}(p, \omega+\psi)\, l\,|p-\varrho\cos\omega|\, dp.$$

Es genügt nun, wenn wir zeigen:

$$(4)\qquad \int_{-\infty}^{+\infty} f(\varrho, 0)\, d\varrho = F\left(0, \frac{\pi}{2}\right),$$

da die Bedingungen a_2—c_2 gegenüber Bewegungen invariant sind. Wir setzen:

$$F(p, \varphi) = F\left(p, \frac{\pi}{2}\right) + \cos\varphi \cdot G(p, \varphi).$$

G genügt leicht angebbaren Regularitätsbedingungen. Es zerfällt nun dieser Zerlegung zufolge $f(\varrho, 0)$ in zwei Bestandteile $f_1(\varrho)$ und $f_2(\varrho)$, die getrennt zu untersuchen sind. Wegen:

$$\int_0^{\pi} l\,|p-\varrho\cos\omega|\, d\omega = \begin{cases} \pi l \dfrac{|p| + \sqrt{p^2-\varrho^2}}{2}, & |p| > |\varrho| \\[2ex] \pi l \dfrac{|\varrho|}{2}, & |p| \leqq |\varrho| \end{cases}$$

ergibt sich:

$$f_1(\varrho) = \frac{1}{2\pi^2}\int_0^{\pi} d\omega \int_{-\infty}^{+\infty} F_{pp}\left(p, \frac{\pi}{2}\right) l\,|p-\varrho\cos\omega|\, dp$$

$$= \frac{1}{2\pi}\int_{|\varrho|}^{\infty} F_{pp}\left(p, \frac{\pi}{2}\right) l\,\frac{|p| + \sqrt{p^2-\varrho^2}}{|\varrho|}\, dp$$

$$+ \frac{1}{2\pi}\int_{-\infty}^{-|\varrho|} F_{pp}\left(p, \frac{\pi}{2}\right) l\,\frac{|p| + \sqrt{p^2-\varrho^2}}{|\varrho|}\, dp.$$

Dies ist nun nach ϱ von $-\infty$ bis $+\infty$ absolut integrabel, wie durch Vertauschung der Integrationsfolge ersichtlich wird. Als Wert des Integrales erhält man:

$$\int_{-\infty}^{+\infty} f_1(\varrho)\, d\varrho = \frac{1}{2\pi}\int_{-\infty}^{+\infty} F_{pp}\left(p, \frac{\pi}{2}\right) \int_{-|p|}^{+|p|} l\,\frac{|p| + \sqrt{p^2-\varrho^2}}{|\varrho|}\, d\varrho$$

$$= \frac{1}{2}\int_{-\infty}^{+\infty} F_{pp}\left(p, \frac{\pi}{2}\right) |p|\, dp = F\left(0, \frac{\pi}{2}\right)$$

Was nun $f_2(\varrho)$ anbelangt, so werden wir zeigen, daß es ebenfalls absolut integrabel ist und von $-\infty$ bis $+\infty$ integriert Null ergibt.

Wir können nämlich $f_2(\varrho)$ folgendermaßen schreiben:

$$f_2(\varrho) = \frac{1}{2\pi^2}\int_0^{\pi} d\omega \int_{-\infty}^{+\infty} G_{pp}(p, \omega)\, l\,|p - \varrho\cos\omega| \cdot \cos\omega\, d\omega$$

$$= \frac{1}{2\pi^2}\int_0^{\pi} d\omega \int_{-\infty}^{+\infty} G_{pp}(p, \omega)\left[l\left|\frac{p-\varrho\cos\omega}{\varrho\cos\omega}\right|\cos\omega + \frac{\varrho p\cos^2\omega}{1+\varrho^2\cos^2\omega}\right]dp,$$

da die hinzugefügten Glieder integriert Null ergeben, und in dieser Gestalt führt die Integration nach ϱ auf ein *absolut* konvergentes dreifaches Integral. Es ist nämlich:

$$\int_{-\infty}^{+\infty}\left|\cos\omega\, l\left|\frac{p-\varrho\cos\omega}{\varrho\cos\omega}\right| + \frac{\varrho p\cos^2\omega}{1+\varrho^2\cos^2\omega}\right| d\varrho$$

$$= |p|\int_{-\infty}^{+\infty}\left|l\left|1 - \frac{1}{\tau}\right| + \frac{p^2\tau}{1+p^2\tau^2}\right| d\tau = \lambda(p)$$

mit

$$\lim_{|p|\to\infty}\frac{\lambda(p)}{|p|\,l\,|p|} = 2.$$

Die Integration nach ϱ liefert als Wert des Integrales:

$$\int_{-\infty}^{+\infty} f_2(\varrho)\,d\varrho = 0,$$

womit (4) bewiesen ist.

Wir haben noch zu zeigen, daß f den Forderungen a_1—c_1 genügt.

Die Stetigkeit folgt aus der Darstellung (3) wegen der Voraussetzungen a_2—c_2. Forderung b_1 ist gleichfalls erfüllt, denn

$$\int_{-\infty}^{+\infty} |f(\varrho\cos\psi, \varrho\sin\psi)|\,d\varrho$$

ist, wie man leicht sieht, nach ψ integrierbar.

Um noch c_1 nachzuweisen, bilden wir:

$$\bar{f}_0(\varrho) = \frac{1}{2\pi}\int_0^{2\pi} f(\varrho\cos\psi,\ \varrho\sin\psi)\,d\varphi$$

$$= \frac{1}{4\pi^3}\int_0^{\pi} d\omega \int_0^{2\pi} d\psi \int_{-\infty}^{+\infty} F_{pp}(p,\psi)\, l\,|p-\varrho\cos\omega|\,dp$$

$$= \frac{1}{4\pi^2}\int_0^{2\pi} d\psi \Bigg[\int_{-\infty}^{-|\varrho|} F_{pp}(p,\psi)\, l\,\frac{|p|+\sqrt{p^2-\varrho^2}}{2}\,dp$$

$$+\int_{|\varrho|}^{+\infty} F_{pp}(p,\psi)\, l\,\frac{|p|+\sqrt{p^2-\varrho^2}}{2}\,dp$$

$$+ F_p(\varrho,\psi)\, l\frac{\varrho}{2} - F_p(\varrho,\psi)\, l\frac{\varrho}{2}\Bigg],$$

woraus man die Richtigkeit von c_1 erkennt. Damit ist Satz IV bewiesen.

B. Bestimmung einer Geradenfunktion aus ihren Punktmittelwerten.

3. $F(p,\varphi) = F(-p,\varphi+\pi)$ sei eine Geradenfunktion, die folgende Regularitätsbedingungen erfülle:

a_3) F, F_φ, F_p seien stetig, $|F_\varphi| < M$ für alle p, φ.

b_3) $F_p \cdot l|p|$ konvergiere für $p \to \infty$ gleichmäßig in φ gegen Null

c_3) $\int_{-\infty}^{+\infty} |F_p|\cdot l\,|p|\,dp$ konvergiere gleichmäßig in φ.

Diese Bedingungen sind wieder gegenüber Bewegungen invariant.

Wir bilden den Punktmittelwert von $F(p,\varphi)$ für $P = [x, y]$:

$$(5)\qquad f(x,y) = \frac{1}{\pi}\int_{-\frac{\pi}{2}}^{+\frac{\pi}{2}} F(x\cos\varphi + y\sin\varphi,\ \varphi)\,d\varphi.$$

Dann gilt:

Satz V: Durch Angabe von f ist F eindeutig bestimmt und zwar ist:

$$F\left(0, \frac{\pi}{2}\right) = -\frac{1}{2\pi}\int_{-\infty}^{+\infty}\frac{dx}{x}\int_{-\infty}^{+\infty} f_x(x, y)\,dy, \tag{V}$$

wo das Integral nach x als CAUCHY*scher Hauptwert zu deuten ist und der Wert von F für irgendeine andere Gerade aus der angeschriebenen Formel durch eine entsprechende Bewegung abgeleitet werden kann.*

Zum Beweise leiten wir aus (5) zunächst ab:

$$\int_{-A}^{B} f_x(x, y)\,dy = \frac{1}{\pi}\int_{-\frac{\pi}{2}}^{+\frac{\pi}{2}} d\varphi \int_{-A}^{B} F_p(x\cos\varphi + y\sin\varphi, \varphi)\cos\varphi\,dy, \tag{6}$$

wo A, B zwei positive Konstante sind.

Wir setzen nun, wie schon früher analog geschehen ist:

$$F(p, \varphi) = F(p, 0) + \sin\varphi\, G(p, \varphi),$$

wobei $G(p, \varphi)$ im Integrationsgebiete beschränkt bleibt und für $p \to \infty$ den Grenzwert Null erhält. Aus:

$$\int_{-A}^{B} G_p(x\cos\varphi + y\sin\varphi, \varphi)\cos\varphi \cdot \sin\varphi\,dy$$

$$= [G(x\cos\varphi + B\sin\varphi, \varphi) - G(x\cos\varphi - A\sin\varphi, \varphi)]\cos\varphi$$

folgt, daß der zweite Bestandteil von (6) für $A \to \infty$, $B \to \infty$ den Grenzwert Null erhält, so daß nur der erste zu untersuchen bleibt. Durch die analoge Integration erkennt man, daß in diesem ersten Bestandteil das Integral nach φ über jedes $\varphi = 0$ *nicht* enthaltende Intervall für $A \to \infty$, $B \to \infty$ ebenfalls Null wird; es bleibt also zu betrachten:

$$\lim_{\substack{A\to\infty \\ B\to\infty}} \frac{1}{\pi}\int_{-\epsilon}^{+\epsilon} d\varphi \int_{-A}^{B} F_p(x\cos\varphi + y\sin\varphi, 0)\cos\varphi\,dy, \qquad 0 < \epsilon < \frac{\pi}{2}.$$

Man kann dieses Integral schreiben:

$$\frac{1}{\pi}\int_{-\varepsilon}^{+\varepsilon} d\varphi \int_{x\cos\varphi - A\sin\varphi}^{x\cos\varphi + B\sin\varphi} F_p(p, 0)\operatorname{ctg}\varphi\, dp$$

und erhält daraus, wenn A und B genügend groß angenommen werden, durch Vertauschung der Integrationsfolge nach einiger Rechnung den Wert:

$$\frac{1}{\pi}\int_{x\cos\varepsilon - B\sin\varepsilon}^{x\cos\varepsilon + B\sin\varepsilon} l\frac{(B^2+x^2)\sin\varepsilon}{|Bp - x\sqrt{B^2+x^2-p^2}|} F_p(p, 0)\, dp$$

$$+\frac{1}{\pi}\int_{x\cos\varepsilon - A\sin\varepsilon}^{x\cos\varepsilon + A\sin\varepsilon} l\frac{(A^2+x^2)\sin\varepsilon}{|Ap - x\sqrt{A^2+x^2-p^2}|} F_p(p, 0)\, dp.$$

Es genügt, den Grenzwert des zweiten Integrales für $A \longrightarrow \infty$ zu ermitteln. Wir schreiben es so:

$$\frac{1}{\pi} l(A\sin\varepsilon)[F(x\cos\varepsilon + A\sin\varepsilon, 0) - F(x\cos\varepsilon - A\sin\varepsilon, 0)]$$

$$+\frac{1}{\pi}\int_{x\cos\varepsilon - A\sin\varepsilon}^{x\cos\varepsilon + A\sin\varepsilon} l\frac{1}{|p-x|} F_p(p, 0)\, dp$$

$$+\frac{1}{\pi}\int_{x\cos\varepsilon - A\sin\varepsilon}^{x\cos\varepsilon + A\sin\varepsilon} l\frac{|Ap + x\sqrt{A^2+x^2-p^2}|}{A|p+x|} F_p(p, 0)\, dp.$$

Da der Logarithmus in dem letzten Integral für $A \longrightarrow \infty$ *gleichmäßig* gegen Null geht, so folgt als Grenzwert:

$$-\frac{1}{\pi}\int_{-\infty}^{+\infty} F_p(p, 0)\, l\,|p-x|\, dp,$$

womit der Grenzwert von (6) erhalten wird:

$$\int_{-\infty}^{+\infty} f_x(x, y)\, dy = -\frac{2}{\pi}\int_{-\infty}^{+\infty} F_p(p, 0)\, l\,|p-x|\, dp.$$

Es mag hier bemerkt werden, daß letzterer Ausdruck die Randwerte des Imaginärteils einer in der oberen Halbebene regulären analytischen Funktion darstellt, für welche die Randwerte des Realteiles den Wert $2F(x, 0)$ haben.

Bilden wir jetzt im Sinne der Formel (V):

$$-\int_{-\infty}^{+\infty}\frac{dx}{x}\int_{-\infty}^{+\infty}f_x(x, y)dy = \frac{2}{\pi}\int_{0}^{\infty}\frac{dx}{x}\int_{-\infty}^{+\infty}F_p(p, 0)\, l\left|\frac{p-x}{p+x}\right|dx,$$

so ist dieses Doppelintegral absolut konvergent und führt wegen

$$\int_{0}^{\infty} l\left|\frac{p-x}{p+x}\right|\frac{dx}{x} = -\frac{\pi^2}{2}\operatorname{sgn} p$$

genau zur Formel (V).

4. Es sei nun f eine Punktfunktion mit folgenden Regularitätseigenschaften:

a_4) f sei mit seinen Ableitungen bis zur zweiten Ordnung einschließlich stetig.

b_4) Die Ausdrücke

$$f(x, y), \quad \sqrt{x^2+y^2}\, l(x^2+y^2) f_x(x, y), \quad \sqrt{x^2+y^2}\, l(x^2+y^2) f_y(x, y)$$

haben für $x^2 + y^2 \longrightarrow \infty$ den Grenzwert Null.

c_4) Die Integrale

$$\int_{-\infty}^{+\infty}\int_{-\infty}^{+\infty} D_1 f \frac{l(x^2+y^2)}{\sqrt{x^2+y^2}}\,dx\,dy \quad \text{und} \quad \int_{-\infty}^{+\infty}\int_{-\infty}^{+\infty} D_2 f\cdot l(x^2+y^2)\,dx\,dy,$$

wo $D_1 f$ jede erste, $D_2 f$ jede zweite Ableitung von f bedeutet, konvergieren absolut.

Diese Bedingungen sind wieder gegenüber Bewegungen invariant.

Dann gilt:

Satz VI: Die aus f nach (V) gebildete Geradenfunktion besitzt die Punktmittelwerte $f(x, y)$.

Es genügt, den Beweis für den Nullpunkt zu erbringen. Für eine beliebige Gerade durch diesen ergibt (V) nach einer partiellen Integration:

$$F(0, \varphi) = \frac{1}{2\pi}\int_{-\infty}^{+\infty}\int_{-\infty}^{+\infty}[f_{xx}\cos^2\varphi + 2 f_{xy}\sin\varphi\cos\varphi$$

$$+ f_{yy}\sin^2\varphi]\, l\,|x\cos\varphi + y\sin\varphi|\,dx\,dy$$

oder nach Einführung von Polarkoordinaten ϱ, ψ:

$$F(0,\varphi)=\frac{1}{2\pi}\int_0^\infty \varrho\,d\varrho\int_0^{2\pi}\Big[\frac{\partial^2 f}{\partial\varrho^2}\cos^2(\varphi-\psi)$$

$$+2\frac{\partial^2 f}{\partial\varrho\,\partial\psi}\frac{\sin(\varphi-\psi)\cos(\varphi-\psi)}{\varrho}+\frac{\partial^2 f}{\partial\psi^2}\frac{\sin^2(\varphi-\psi)}{\varrho^2}$$

$$+\frac{\partial f}{\partial\varrho}\frac{\sin^2(\varphi-\psi)}{\varrho}$$

$$-2\frac{\partial f}{\partial\psi}\frac{\sin(\varphi-\psi)\cos(\varphi-\psi)}{\varrho^2}\Big]\,l\,|\varrho\cos(\varphi-\psi)|\,d\psi.$$

Um den Punktmittelwert für $[0, 0]$ zu bilden, kann man die Integration nach φ unter dem Doppelintegral von 0 bis 2π ausführen und hat dann noch durch 2π zu dividieren. Das Glied mit $\frac{\partial^2 f}{\partial\psi^2}$, das dabei zum Vorschein kommt, fällt vermöge Integration nach ψ weg und es bleibt:

$$\frac{1}{2\pi}\int_0^{2\pi}d\psi\int_0^\infty\Big[\frac{1}{4}\Big(\varrho\frac{\partial^2 f}{\partial\varrho^2}-\frac{\partial f}{\partial\varrho}\Big)+\frac{1}{2}l\frac{\varrho}{2}\frac{\partial}{\partial\varrho}\Big(\varrho\frac{\partial f}{\partial\varrho}\Big)\Big]d\varrho,$$

was sich in der Tat auf $f(0, 0)$ reduziert.

Um die eindeutige Bestimmtheit von F zu zeigen, wären die Bedingungen a_3—c_3 als erfüllt nachzuweisen, was offenbar noch weitere Voraussetzungen über f nötig macht.

5. Es möge hier folgende Bemerkung Platz finden, die ich, wie überhaupt die Problemstellung B, Herrn W. BLASCHKE verdanke: Die beiden hier behandelten Aufgaben stehen in engem Zusammenhange mit der Theorie des NEWTONschen Potentials. Betrachten wir nämlich den Übergang von einer Punktfunktion f zu ihren geradlinigen Mittelwerten F als eine lineare Funktionaltransformation

$$F = Rf,$$

und ebenso den Übergang von einer Geradenfunktion F zu ihren Punktmittelwerten v:

$$v = BF,$$

so liegt es nahe, die zusammengesetzte, durch

$$v = Hf = B[Rf] = BRf$$

definierte Transformation $H = BR$ zu betrachten.

Man sieht nun sofort, daß Hf nichts anderes ist, als das NEWTONsche Potential der mit der Massendichte $\frac{1}{\pi}f$ belegten Ebene in den Punkten der Ebene selbst. Daraus kann man nach einer Bemerkung von G. HERGLOTZ die Umkehrung der Transformation H gewinnen; es ergibt sich dabei:

$$f(P) = H^{-1}v = -\frac{1}{2}\int_0^\infty \frac{d\bar{v}_P(r)}{r} = -\frac{1}{4\pi}\int_{-\infty}^{+\infty}\int_{-\infty}^{+\infty} \frac{\Delta v(x'y')}{r_{PP'}}\,dx'dy',$$

wo $\bar{v}_P$ eine zu den früher eingeführten analoge Mittelwertsbezeichnung ist und Δ den LAPLACEschen Operator bedeutet.

Nun liegt der Gedanke nahe, die in 1.—3. direkt geleitete Umkehrung von R und H durch den Ansatz

$$R^{-1} = H^{-1}B \quad \text{bzw.} \quad B^{-1} = RH^{-1}$$

zu leisten. Tatsächlich habe ich die Umkehrungsformel (IV) zuerst auf diesem Wege gefunden, aber eine strenge Durchführung dieses Gedankens scheint schwieriger zu sein als die direkte Verifikation und versagt auch in den gleich zu besprechenden nichteuklidischen Fällen.

Schließlich sei bemerkt, daß die in A und B zugrunde gelegten Regularitätsbedingungen selbstredend bei weitem nicht die allgemeinsten sind, wie sich an einfachen Beispielen zeigen läßt.

C. Verallgemeinerungen.

6. Eine weitgehende Verallgemeinerung des in A behandelten Problems ließe sich etwa folgendermaßen formulieren: es sei eine Fläche S gegeben, auf der irgendwie ein Bogendifferential ds definiert sei, ferner eine zweifach unendliche Schar von Kurven C auf S. Es soll eine Punktfunktion der Fläche aus ihren Integralwerten $\int f ds$ längs der Kurven C bestimmt werden.

Die nächstliegende Spezialisierung erhält man, wenn man für S eine nichteuklidische Ebene, für ds das zugehörige Bogenelement und für die Kurven C die Geraden nimmt. Im elliptischen Falle kann man die Aufgabe auf die Kugelgeometrie hinüberspielen; indem man in bekannter Weise ein diametrales Punktepaar der Kugel als Punkt der elliptischen Ebene deutet, erhält man die Aufgabe, eine gerade — d. h. in diametralen Punkten gleichwertige — Funktion auf der Kugel aus ihren Großkreisintegralen zu bestimmen. MINKOWSKI hat diese Aufgabe im Prinzip zuerst behandelt[1]) und durch Entwicklung nach Kugelfunktionen gelöst; P. FUNK hat später die MINKOWSKIsche Lösung durchgeführt und gezeigt, wie man die Lösung mit Hilfe der ABELschen Integralgleichung finden kann[2]) und dieser Methode verdanke auch ich die Lösung des Problems A. Die FUNKsche Lösung ist ganz analog zu (III), nur tritt in den Nenner der Sinus des sphärischen Radius und additiv zu dem Integral der Wert von F im Pole des betreffenden Großkreises dividiert durch π. Aber auch in der hyperbolischen Ebene hat die gestellte Aufgabe die zu (III) analoge Lösung:

$$f(P) = -\frac{1}{\pi}\int_0^\infty \frac{d\overline{F}_P(q)}{\mathfrak{Sin}\, q}$$

(es ist hier das Krümmungsmaß $= -1$ angenommen), wie sich ganz konform zu der in 1. angedeuteten Ableitung von (III) zeigen läßt.

In beiden Fällen kann man auch die zu B. analoge Frage stellen. In der elliptischen Geometrie erhält man vermöge der absoluten Polarität nichts Neues, im hyperbolischen Falle scheint eine zu (V) analoge Lösung nicht zu existieren.

Eine zweite Spezialisierung ergibt sich, wenn man (in der euklidischen oder nichteuklidischen Geometrie) als Kurven C die Kreise mit konstantem Radius nimmt. Hier kann man auf der Kugel die MINKOWSKIsche Behandlung mit Kugelfunktionen anwenden und die Aufgabe in gewissem Grade lösen. Interessant ist in diesem Falle, daß die Eindeutigkeit der Lösung verloren gehen kann; es gibt nämlich für gewisse, durch die Nullstellen der LEGENDREschen Polynome gerader Ordnung definierte Radien ϱ gerade Funktionen auf der Kugel, die längs jedes Kreises vom sphärischen Radius ϱ integriert, Null ergeben, ohne identisch zu verschwinden. Im euklidischen Falle tritt an Stelle der Kugelfunktionenreihen das Integraltheorem der Besselfunktionen; hier gibt es stets Funktionen, die über alle Kreise von festem Radius integriert, Null geben und doch nicht identisch verschwinden; ist der Radius 1, so sind es (in Polarkoordinaten ϱ, φ) die Funktionen

$$J_n(x_\nu \varrho)\cos n\varphi, \quad J_n(x_\nu \varrho)\sin n\varphi,$$

und ihre Linearkombinationen, wo x_ν eine Nullstelle von J_0 ist. Im hyperbolischen Falle treten an Stelle der Besselfunktionen die

1) Ges. Abh., Bd. II, S. 277f. 2) Math. Ann., Bd. 74, S. 283—288.

sogenannten Kegelfunktionen, für die das zugehörige Integraltheorem von WEYL[1]) bewiesen ist. Die Ergebnisse sind analog zum euklidischen Falle.

7. In anderer Richtung verallgemeinern sich die Ergebnisse von A und B beim Übergang zu höheren Räumen. In einem euklidischen R_n' kann man eine Punktfunktion $f(P)=f(x_1, x_2, \ldots x_n)$. aus ihren Integralwerten $F(\alpha_1 \ldots \alpha_n, p)$ über alle Hyperebenen $\alpha_1 x_1 + \cdots + \alpha_n x_n = p$ $(\alpha_1^2 + \cdots + \alpha_n^2 = 1)$ zu bestimmen trachten. Analog zu dem in 1. eingehaltenen Vorgang bilden wir den Mittelwert $\overline{F}_0(q)$ von F über die Tangentialebenen der Kugel vom Zentrum $[0, 0 \ldots 0]$ und Radius q. Er ist durch das $n - 1$ fache Integral:

$$\overline{F}_0(q) = \frac{1}{\Omega_n} \int F(\alpha, q)\, d\omega$$

gegeben, wo $d\omega$ das Oberflächenelement, $\Omega_n = \dfrac{2\pi^{\frac{n}{2}}}{\Gamma\left(\frac{n}{2}\right)}$ die Oberfläche der n-dimensionalen Kugel $\alpha_1^2 + \cdots + \alpha_n^2 = 1$ bedeutet.

Man kann $\overline{F}_0$ durch ein n-faches Integral über f darstellen und zwar ergibt sich:

$$(7) \quad \overline{F}_0(q) = \frac{\Omega_{n-1}}{\Omega_n} \int \cdots \int\limits_{x_1^2 + \cdots + x_n^2 > q^2} f(x_1, x_2, \ldots x_n) \frac{(x_1^2 + \cdots + x_n^2 - q^2)^{\frac{n-3}{2}}}{(x_1^2 + \cdots + x_n^2)^{\frac{n-2}{2}}}\, dx_1 \ldots dx_n$$

oder in einer nun schon oft verwendeten Mittelwertsbezeichnung:

$$\overline{F}_0(q) = \Omega_{n-1} \int_q^\infty \overline{f}_0(r)(r^2 - q^2)^{\frac{n-3}{2}}\, r\, dr.$$

Dies ist die zu (1) analoge Formel, an die sich entsprechende Folgerungen anschließen. Die Substitution $r^2 = v$, $q^2 = u$ führt auf die Integralgleichung:

$$\Phi(u) = \frac{\Omega_{n-1}}{2} \int_u^\infty \varphi(v)(v-u)^{\frac{n-3}{2}}\, dv.$$

1) Gött. Nachr. 1910, S. 454.

Ist n *gerade*, so ergibt $\left(\frac{n}{2} - 1\right)$-maliges Differentiieren nach u dieselbe Gleichung wie (2) und man kann hieraus

$$\varphi(0) = f(0, 0, \dots 0)$$

finden, wozu also bei gegebenem F die Bildung von $\overline{F}$, Differentiationen und eine Integraloperation nötig sind. Bei *ungeradem* n fällt diese Integraloperation weg, denn jetzt ergibt $\left(\frac{n-1}{2}\right)$-maliges Differentiieren:

$$\varphi(0) = \frac{2(-1)^{\frac{n-1}{2}}}{\Omega_{n-1}\left(\frac{n-3}{2}\right)!} \Phi^{\left(\frac{n-3}{2}\right)}(0).$$

Besonders einfach gestaltet sich der dreidimensionale Fall; diesen kann man aber auch nach einer Methode behandeln, die zu 5. analog ist und sehr elegante Ergebnisse liefert. Aus (7) geht nämlich für $q = 0$ der Punktmittelwert von F hervor:

$$\overline{F}_0 = \frac{1}{2} \iiint \frac{f(x, y, z)}{\sqrt{x^2 + y^2 + z^2}} \, dx \, dy \, dz,$$

der als das NEWTONsche Potential des mit der Massendichte $\frac{1}{2} f$ belegten Raumes anzusprechen ist. Also folgt:

$$f(x, y, z) = -\frac{1}{2\pi} \Delta \overline{F},$$

wo $\overline{F}$ den Punktmittelwert von F bedeutet.

Hier kann man auch die zu B. analoge Aufgabe lösen und findet nach der in 5. angedeuteten Methode für eine Ebenenfunktion F, deren Punktmittelwerte f bekannt sind:

$$F(E) = -\frac{1}{2\pi} \iint \Delta f \, d\sigma,$$

wo $d\sigma$ das Flächenelement der Ebene E ist. Δ ist der LAPLACEsche Operator für den dreidimensionalen Raum, die Integration über die ganze Ebene E zu erstrecken.

Druckfertig erklärt 1. X. 1917.]

ABCDEFGHIJ–CM–89876543